PLANNING
LOCAL
ECONOMIC
DEVELOPMENT

PLANNING LOCAL ECONOMIC DEVELOPMENT

Theory and Practice

Second Edition

Edward J. Blakely

SAGE Publications
International Educational and Professional Publisher
Thousand Oaks   London   New Delhi

For information address:

 SAGE Publications, Inc.
2455 Teller Road
Thousand Oaks, California 91320

SAGE Publications Ltd.
6 Bonhill Street
London EC2A 4PU
United Kingdom

SAGE Publications India Pvt. Ltd.
M-32 Market
Greater Kailash I
New Delhi 110 048 India

Printed in the United States of America

Library of Congress Cataloging-in-Publication Data

Blakely, Edward James, 1938-
    Planning local economic development: theory and practice/Edward
J. Blakely. — 2nd ed.
        p. cm.
    Includes bibliographical references and index.
    ISBN 0-8039-5209-0.–ISBN 0-8039-5210-4 (pbk.)
    1. Industrial promotion–United States. 2. Community development–
United States. 3. Economic development. I. Title.
HC110.I53B56   1994
338.973–dc20   93-37618
                                                              CIP

94 95 96 97 98 10 9 8 7 6 5 4 3 2 1

Sage Production Editor: Yvonne Könneker

Contents

Acknowledgments

Like all scholar-practitioners, I am most grateful to the people in the communities where I have worked and learned. They have enriched this second edition of a popular book. Although success usually breeds imitators, this did not happen after the first printing. There have been a number of new works on local economic development, but none has been written as a competitive text. Most of the new books and articles supplement the first edition of this book; several, in fact, have been published by Sage. These recent publications and the increased circulation of *Economic Development Quarterly* have, however, stimulated interest in the field as well as created a greater demand for this second edition. I am deeply grateful to my colleagues for the new ideas and material that they have produced and that I have tried to incorporate. Obviously, I have not been able to incorporate everything produced since 1989.

I had considerable assistance in various ways in putting this new edition together. Colleagues around the world have written me offering comments and criticisms. I am grateful to them. I have had additional opportunities to experience economic development firsthand in Asia, Europe, and the United States. I am particularly grateful to friends and colleagues in Australia and the United States for giving me many new opportunities to visit, speak, and work in their communities and see economic development firsthand. I am particularly grateful to the people of Oakland, California, for their patience and generosity in allowing me to work with them on economic development projects to revitalize that city. This has been both rewarding and heartbreaking. Any major city has its problems, but Oakland has had more than its share. In 1989, Oakland experienced a devastating earthquake that destroyed its downtown area and crippled its transportation systems. Only 2 years later in 1991, the worst residential fire in the nation's history hit the city. After both events, the mayor and council asked me to assist in developing recovery plans. I accepted these challenges from two different mayors and councils. The City of Oakland is now undergoing significant transformation. In 1993, Oakland was designated as an All American City. I hope what I have contributed has made some small difference. Oakland has made economic development very real for me and has been an experiment that has shaped both this book and my life.

Edited by Eleanor Shapiro, this second edition includes a new section of illustrative case materials. I have continued composite case studies because of their instructive nature. The case studies are useful because they present instructive materials that add to each chapter.

I am indebted to my colleagues Mike Teitz and Ted Bradshaw for helping me through the process of co-teaching much of the material in this book. I am also grateful to Robert Meir of the University of Illinois at Chicago for his guidance and critiques of the earlier work as well as Nancy Green Leigh of the University of Wisconsin-Milwaukee for her clear and careful criticisms. I bear total responsibility for anything included or excluded.

EDWARD J. BLAKELY

Introduction

Local governments and community groups throughout the United States are examining their opportunities for improving the economic and employment base. The Clinton administration has given new impetus to local government initiatives. The administration is backing a broad range of activities, from sponsoring local employment projects to actually acting as joint venture partners with private enterprise to develop and manage new enterprises. Similarly, depressed neighborhoods and rural unincorporated areas are developing their own approaches to stimulating local business. They are creating new institutions such as cooperatives to generate jobs or stabilize the community's economic base. In some instances, local governments and groups are acting on their own; in others, they are cooperating with other communities or tiers of government to embark on economic development activities. Localities have selected organizational structures and initiatives for development activities nearly as diverse as the nation itself. Irrespective of the form, the goal is the same; namely, to provide the neighborhood, locality, or region with greater capacity to contribute to determining its own economic destiny.

Local institutional involvement in economic development is an extremely complex process. It requires increased knowledge and cooperation on the part of government officials, business, unions, professional leaders, and community groups. Therefore, community organizations or local government officials must carefully consider whether they possess the necessary institutional resources as well as economic development opportunities before they embark on any activities. Economic development may not be useful or reasonable for every situation. Nonetheless, communities irrespective of their size or location must consider their economic destiny as a major component of their political agenda. One thing is very clear: The forces of economic change do not respect national, regional, local, or community boundaries.

This second edition of *Planning Local Economic Development* is aimed primarily at students studying to become professional practitioners of local economic development at the regional agency, city or county government, and neighborhood levels. It is hoped that it will continue to be a reference for professional practitioners and economic development or planning specialists as they carry out their responsibilities. Economic development specialists and planners will find the resource materials included in this book useful for their work. Moreover, they may find the examples or illustrative economic development instruments (tools) particularly helpful in designing various local projects.

In addition, this edition continues to provide guidance for community-based organizations and their client groups engaged in the struggle to find viable means to develop the economic base of their communities. The book is also designed to serve as primary material or as an adjunct to texts or other materials used in teaching economic development. Both the reference materials and the case examples are arranged in a manner suitable for teaching. Sample resource materials are included to facilitate its use in this manner. The sample materials are also intended to support analytical efforts designed to determine community economic and employment requirements, as well as to develop project or program ideas.

Policymakers at either the local government or the community level may find the book useful in exploring the role that private enterprise, unions, community groups, and other institutions can play in local economic and employment development activities. In addition, it can aid private citizens interested in local economic development and employment issues to determine ways they might assist government bodies or community groups to meet their economic development and/or employment objectives. Community groups may find the illustrations and the case examples discussed throughout the text very valuable in forming networks and making contact with people in similar circumstances across the nation.

Local economic development and employment initiatives are a new field of endeavor. As such, there are no hard rules nor is there long experience on which to draw. This book is based on the needs, issues, and options available at the local level. I hope that its readers will contribute to future editions.

This edition is based on my continued field experience throughout the United States, Asia, and in my hometown of Oakland, California. I also continue to work with practitioners of local economic development in rural areas throughout the United States. This work requires me to be very specific with respect to the means and aims of economic development lest that term be used as another vehicle to stifle and urbanize rural America.

No new book has been written that places all the tenets of economic development in one volume. Much of what is written about local economic development is cast in a real estate and factory-chasing guise, with little relevance to those who believe in local development as a "community-determined process." The frustration of this situation led me to consider the possibility of writing a book on local economic development based on the courses I offer in this field with a focus on urban communities.

In recent years, the quality and availability of material in the field of economic development has improved considerably, particularly since *Economic Development Quarterly*, also published by Sage, commenced publication in 1985. In addition, the American Planning Association has published articles in its journal and monographs, through its press, on various aspects of local economic development. Although this book includes case studies, they are not meant to replace the current literature offered by these publications and many others that are emerging. Rather, the book is a condensed statement and reference guide that provides a coherent body of knowledge in order to frame the topics of discussion in this growing field.

Terms Used in this Book

Local economic development and *employment generation* are relatively new terms that have recently come to mean a number of things depending on where they are used. There are no authoritative definitions for these and other terms used to describe the activities being undertaken around the world to stimulate local economic activity and employment. I will attempt to be consistent, however, so that their meanings are clear, at least in the context in which I am using them. Specific terms used regularly are defined here for the reader's convenience:

Local economic development refers to the process in which local governments or community-based (neighborhood) organizations engage to stimulate or maintain business activity and/or employment. The principal goal of local economic development is to stimulate local employment opportunities in sectors that improve the community, using existing human, natural, and institutional resources.

Regional and *local* are used interchangeably to refer to a geographic area composed of a group of local government authorities that generally share a common economic base and are close enough together to allow residents to commute between them for employment, recreation, or retail shopping.

Local government refers to municipal-level government irrespective of whether it is a city, town, county, or municipal corporation.

Employment/job generation or *creation* is used to describe all of the activities in which locally based organizations (public or private) engage to increase employment opportunities in a specific community or area, especially for the disadvantaged and unemployed.

Planning encompasses the broad concept of formulating courses of action for socioeconomic change. It is not used with the restrictive city planning connotation.

Initiatives are purposeful acts undertaken by government, business, unions, and community groups (usually in concert with one another) to achieve desired employment and economic outcomes in a designated geographic area.

Many other terms used throughout the text, such as *strategies*, *tools*, *methods*, and *viability*, are defined by the context in which they are used. A glossary of terms frequently used in connection with economic development is included in the appendix.

This book is arranged in a manner that will allow readers to select from the menu of ideas presented. The assumption is that some readers will know a considerable amount about this topic from their own experience and others from their reading. The case studies incorporate queries at the end of each of them. These may be used to start discussions, as student study guides, or as issues for consideration as a community embarks on similar programs.

Each chapter is written to stand on its own in some respects. However, the first three chapters explain how and why the material included is organized in this manner.

1 | The Argument for Taking Local Economic Development Initiatives

There is increasing national agreement that the experiments of the 1980s, variously labeled as *supply-side*, *free market*, *export-based* and the like, have transformed and weakened local and regional economies. As a result, Americans in every community, from suburbs to rural areas to inner cities, are worried about their economic future. Moreover, state and local officials appear to be helpless in altering the economic fortunes of their jurisdictions. Millions of traditional manufacturing jobs have moved overseas in less than two decades. The strong, safe, and economically mobile middle-class opportunities available to nearly every hardworking and aspiring American for more than 50 years have eroded. There is little doubt that national growth and productivity have slowed. Even working Americans have suffered declines in their living standard. This is partially reflected in a 3.3% decrease in average real hourly wages between 1979 and 1987, which reflects low national economic expansion. Over the past two decades, the nation's productivity has not only declined but fallen relative to other Western industrialized nations. Although the United States continues to lead the world in productivity on the basis of output per hour, this is primarily because of a combination of technology improvements and labor shedding.

American families caught in this downward spiral have tried to keep pace with the nation's changing fortunes by sending more members into the workforce. Many families are stressed and feel insecure about their future. This insecurity is shown in the nation's social life, reflected in deepening racial and socioeconomic divisions (Goldsmith and Blakely 1992). As my coauthor and I state forcefully in a companion book:

1

> The American Century is over. The great postwar boom has col-
> lapsed, and international events have transformed the nation's economy
> and politics. Traditions of social relations are disintegrating. In the
> mid-twentieth century there was a common belief that all Americans
> shared a common economic destiny. The wealth of the nation would
> flow to all citizens who displayed diligence and thrift. This belief
> lasted forty years. After four decades of [economic] progress, the
> basic social contract that connects people and opportunities has
> begun to break. (Goldsmith and Blakely 1992, 15)

The fraying of the social contract was dramatically illustrated in
the post-Rodney King riots of April and May 1992. These economic
riots emerged out of the manifested social despair and compounded
frustrations of people in every racial and income group. Many of the
rioters openly stated that they felt left out and cheated by the
nation's social and economic promises. The promise of a job and
economic security are the hallmarks of citizenship. Work is the basis
of social and economic status. The absence of work or the lack of
opportunity to work destroys the basic building block of the nation's
sociopolitical system. As a result, the nation must re-establish the
opportunity structure by providing every citizen with the interests,
skills, and capacity to contribute through work and community
service. Although the symptoms of joblessness in inner-city Chicago
or Detroit are markedly different from those of small towns in Iowa,
Oregon, South Dakota, the rural South, or the Midwest, the despair
of the communities and their residents is very similar.

Local communities and individual citizens feel increasingly powerless over their own economic destiny when plants close or
reduce their workforces. Regardless of whether local job loss is
called "economic dislocation," "de-industrialization," "market failures," or "*capital mobility*," it is a disaster for the locality where
it occurs—a small town, a major city, or an inner-city neighborhood.
Local civic leaders must mobilize their resources in order to create
alternative economic and employment opportunities. In this context, the concept of *local economic development* is promoted by
its adherents and discussed in this book.

The Forces Shaping Employment Options
and Opportunities in American Communities

Economic stagnation and employment instability have been the
patterns of domestic economy in the 1980s. Worker turnover in the

United States has been over 4% a month, whereas our largest industrial competitor, Japan, has only *a 3.5% turnover annually!* We are now a nation of the unemployed or underemployed. Both blue- and white-collar workers have been affected by the economic malaise. Of nearly 5.1 million full-time workers laid off in 1987, only 2.1 million of them found jobs at or above their former wage rates. Millions of Americans are working at jobs well below their educational capacity, and it is estimated that the nation will produce 30% fewer jobs for college graduates than needed over the 1990-2000 decade. As a result, college graduates are now competing with high school graduates for clerical, sales, and lower-skilled employment. For the first time since the Great Depression, the nation's real unemployment rate approached nearly 15% in 1982 and remains at historically high levels (see Figure 1.1).

These statistics are unsettling for a nation that had the world's fastest growing economy, an economy that expanded by over $2.3 trillion from 1929 to 1990 and produced over 30 million new, good jobs between 1950 and 1970. In the four decades after the second world war, the United States experienced one of the world's longest and most productive economic cycles almost without interruption. The workforce increased in every sector, excluding agriculture and natural resources. Manufacturing employment improve- ments and productivity were spectacular by any standard, providing a degree of economic stimulus for all but the most isolated and underdeveloped areas of the nation.

The job base expanded by 48%, providing more opportunities for women, teenagers, and minorities to enter the labor market. The United States was the leading and best producer of jobs in the world—particularly "good" jobs in the expanding producer services areas of finance, insurance, computing, and related industries. The nation was not only producing "thinking jobs" for white-collar workers but also expanding its skilled manual employment—albeit not at rates comparable to the rapid increase in the service sector. Over 70% of the nation's private employment now lies in service or non-goods-producing areas.

Even when the nation was producing jobs, however, there was not a perfect match between people seeking work, the location of work, and the resources needed to stimulate job creation. Some areas, particularly rural communities and inner-city neighborhoods, lagged well behind the nation in employment-generation pattern. In fact, almost 60% of all the jobs in the nation are now in the suburbs. This movement of jobs from the cities to the suburbs has basically

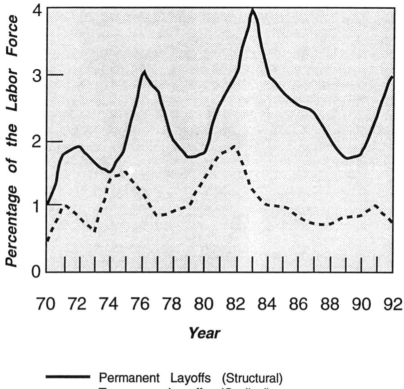

Figure 1.1. Cyclical and Structural Job Losses, 1970-1992

decanted the cities. Cities such as Detroit lost over 600,000 people
between 1960 and 1990; Chicago lost 7.4% and Newark 16% of
their populations in the 1980-1990 decade. These losses were
mirrored by both increased racial and job spatial segregation. Many
leading economists viewed the problem of job movement as the
direct or indirect result of such factors as global market demands,
obsolete plants, uncompetitive industries, locational disadvantages,
inferior skills, and racial discrimination. According to some of
these economists, correct national economic policies could cure
all these problems by stimulating better human and financial capital flows. In their view, unemployment is a problem linked to the

competitive position of the *place/locality* in the national and world economy (Reich 1991). However, even when the economic circumstances of some places are altered, the predicaments of many individuals in these locations are not materially affected.

As the Department of Commerce observed as early as 1974:

> The basic problem . . . [is] that many of the central city and [rural] . . . areas have been for some time experiencing long-term economic decline. Both jobs and workers were involved in a . . . flight to the suburbs. Moreover, the eroding . . . city tax base is shifting the tax burden on those who remain behind. The lack of a skilled labor force, the burden of local taxes, the deteriorating conditions and obsolescence of public and private capital stock, and the re-location of suppliers and markets have contributed to further discouraging new firms and encouraging the departure of existing firms. (U.S. Department of Commerce 1977, 2)

The nation has not embarked on any set of programs or policies with an explicit local economic development orientation. As a result, national policy continues to drift, and the burden of economic revitalization and industrial policy has been placed on the shoulders of local and state officials. These policymakers have not engaged in the national economic policy debates. In contrast, subnational policymakers who have little capacity to directly intervene in a new economic structure or a global marketplace have attempted to structure their own solutions to international forces. It is important that human resources, natural resources, and world market potentials be aligned. That is, if new jobs are to be formed, the jobs need to "fit both the people and the place." Therefore, regional (state and substate) planners must develop economic development strategies that utilize the local resource base (both human and physical) to compete. As Belden Daniels and Lawrence Litvak observe:

> Regardless of how it manifests itself, the existence of relatively depressed communities in substate regions means a certain segment of the population is cut off from the fruits of national economic development. People in these localities will not simply migrate to healthier areas. On the contrary, better educated people with more promising job prospects are likely to move from place to place looking for employment. Moves by poor people tend to be within the same county or city. *Clearly there is a need to try to bring jobs to people rather than counting on people to move to jobs.* (Litvak and Daniels 1979, 1; emphasis added)

Creating new jobs or developing human capacity is not an easy task. Local institutions have little authority and few resources to embark on economic planning.

Economic development as a people-and-place strategy is not new. During the 1960s, the federal government enticed many local governments to embark on a set of social and economic development experiments in an attempt to aid communities and individuals who were not benefitting from economic expansion. A renaissance in neighborhood and central-city revitalization was proposed through the Model Cities Program along with a new urban focus of the Economic Development Administration. (The Economic Development Administration had previously served a rural constituency.) Urban renewal activities as well as other government programs provided local governments with the first impetus to plan for the local economy in a systematic way.

Finally, industrial promotion programs that began in the South in the early 1950s to industrialize that region spread to almost all localities in the nation. The industrial attraction concept and methods were viewed positively by civic leaders during the long wave of economic growth, because there were generally enough new firms and growth for all communities to share. Although spirited competition existed among different localities, the general view was that the economy was dynamic enough to accommodate any reasonable bidder. These initial steps to cure the needs of some disenfranchised groups and places in the 1960s formed the building blocks of a new national effort to revitalize the national economy from the bottom up.

The Causes of the Changed Circumstances

The good times (1950-1974) produced a psychology regarding the nation's capacity to produce jobs and local governments' ability to promote local solutions to imbalances and dislocations in the labor market system. The American economy emerged at the end of the 1980s decade with a combination of new strengths and deep weaknesses. In the 1980s, the United States became an export nation. Expanding exports cut nearly 40% off the national trade deficit. But during the same period, the nation became the world's largest debtor. This debt is troubling in several respects: It is eroding the nation's productive capacity by keeping interest rates high, redirecting private sector investment, and imperiling the future of subsequent generations. The federal deficit in 1992 accounted for nearly 40% of the national budget, and it could rise to 65% of GDP in 2000.

The problems of debt and productivity decline hit the nation in the same year. Until the oil embargo of 1974, the nation, its communities, and its citizens felt that they had greater ability to handle the local problems at hand than they actually possessed. Every community felt that it could lure new firms or depend on local firms to increase production. The subsequent restructuring of the international economy produced a wave of depressed GDP and employment in nearly all industrialized nations, except Japan (see Figure 1.2).

In responding to these conditions, local governments and community leaders attempted to cope with internationally rooted employment problems much broader than their jurisdictions. Local capacity to address economic issues remains a legitimate question. Nonetheless, local leaders have few alternatives. Citizens look to local government, community enterprises, and civic leaders to provide both the leadership and the resources necessary for their employment needs. Local neighborhoods are not powerless to act. They do have resources that can be mobilized to stimulate both improved economic performance and increased employment.

To some extent, local economic health is dependent on national economic performance. For example, the national economic recovery that began in 1985 faltered quickly, and the nation has been in a deeP & Long slump that may last until the middle of the 1990s. Even California, which escaped earlier recession, has been hard hit by the loss of more than 600,000 firms as well as an annual exodus of nearly 300,000 of its most able residents. New England,the economic miracle of the 1980s, has not recovered, and the Southwest and Deep South have slowed their employment growth. Even the most rapidly growing areas have experienced high rates of unemployment for some segments of the community, especially African-Americans, Hispanics, and recent Asian immigrants. Even when manufacturing output expanded in the 1990s, employment in the sector fell by nearly 3.0%, or more than 500,000 workers.

No sector or community is immune from massive economic disruption and unemployment. Joblessness is a concern in all areas of the nation. The previously rapidly growing electronic industry shed 75,000 jobs in 1990 alone. In the so-called Sun Belt of the South, Southwest, and West, the defense industries have shown

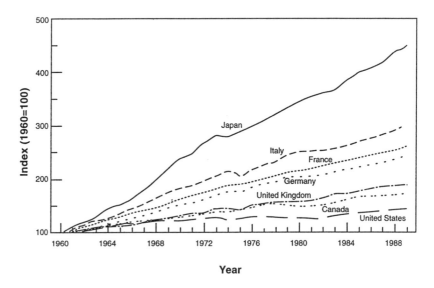

Figure 1.2. Trends in Real GDP per Employed Person, Selected Countries, 1960-1989

SOURCES: Federal Reserve Bank of Boston, Bureau of Labor Statistics.

themselves not to be recession-proof. Durable manufacturing employment—which stood at over 30 million in the 1970s—has fallen to under 18 million in 1992. Mobil and General Electric each decreased their manufacturing employment by more than 100,000 workers in the 1980s. The rising tide of unemployment has created a natural/frictional unemployment rate of over 6%, a level viewed less than a decade ago as twice the rate of total unemployment the nation could reasonably sustain without economic collapse. Some economists predict the natural/frictional unemployment rate will reach 9% by the turn of the century. There is considerable debate as to how and why these circumstances arose and whether they will persist. This debate will not be repeated here. It is sufficient to say that policymakers comprehend the need for continuing economic development activities to produce employment for all segments of the community.

Manufacturing is an area of both employment decline and disinvestment. Closure of manufacturing plants resulting from capital flight, loss of competitiveness, and technological change have reduced worker confidence and that of their communities in the possibility of reestablishing a sense of community prosperity. The nation's Fortune 500 largest firms are the bedrock of many communities. They are restructuring the most rapidly, shedding labor to maintain profits and competitiveness (see Figure 1.3).

Slower Economic Growth

Increases in the nation's overall economic capacity have been an important factor in building local business confidence. Confidence in the economy translates into taking business risks—hiring new employees, adding to product lines, improving services, and increasing advertising. All of these activities slow down when the national economy appears to be sluggish, misguided, or in recession. The nation's GNP and productivity are not as robust as they were in the 1960s. For example, total GNP increases fell from 4.2% in the 1960s to 2.7% in 1984 and 2.0% in 1992. Manufacturing productivity figures in the same period provide a mixed record, showing growth in person-hour output but a decline in the number of personnel used. Median weekly wages in manufacturing fell from $430.30 in 1979 to $415.43 in 1989. Further, a new threatening phenomenon known as "jobless" growth has surfaced in manufacturing and goods-producing sectors. This new growth pattern refers to improvements in productivity without corresponding increases in human resources.

Regional and Metropolitan/Nonmetropolitan Shifts in Employment

Jobs have been far more mobile than people. In the last three decades, the Northeast and Midwest "Snow/Rust Belts" have shed manufacturing and related jobs. In contrast, the South and West have added defense, construction, and electronics employment. Southern metropolitan areas added manufacturing jobs in the 1970s, while the North lost these same jobs during that period.

The movement of jobs has not been confined to north-south or east-west shifts. Northern metropolitan regions lost population and employment during the same decade to nearby small towns and suburbs. Rural population increased in the 1960s and 1970s, along with corresponding employment increases in manufacturing, trade, and professional services. Rural communities in favored resort and good weather locations, particularly in the West, found that their major asset was the lifestyle amenity their relative isolation from the metropolis afforded them (Blakely and Bradshaw 1986; Pigg

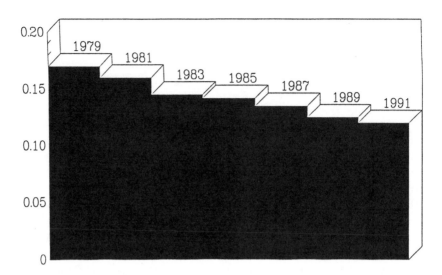

Figure 1.3. Fortune 500 Companies Employment (millions), 1979-1991
SOURCE: Pacific Gas and Electric, 1992.

1991). Moreover, the skill level of rural dwellers has now increased to a degree competitive with urban and suburban areas.

International Competition

The U.S. economy has become increasingly internationalized. American trade increased from approximately $15 billion in 1950 to over $190 billion in 1990, or almost 30% of all goods purchases in the nation. Increased productivity and competitiveness of our major trading partners combined with the rising trading capacity of the new free market nations, formerly central-planned economies in Eastern Europe and China, directly affect U.S. workers. As U.S. firms enter this competition (and, some would argue, sponsor it), their loyalty to national boundaries diminishes. By 1989, for example, 46% of IBM's assets and 44% of its employees were located outside of the United States. American direct investment overseas jumped from $215 million per year in 1980 to $373 million in 1988, most of which was directly invested in manufacturing activity in competition with the American workforce. Many U.S. firms are therefore engaged in what has come to be known as "exporting jobs." That is, firms seek the best location to obtain the good, service, or workforce necessary for production without adhering to national boundaries.

In the early 1990s, this trend was especially noticeable as large manufacturing firms such as Phillips Corp. and AT&T moved thousands of $13-an-hour jobs to Mexico, where they pay $3 an hour. The auto industry has also been busily closing down plants and exporting jobs. By the year 2000, the three major U.S. auto companies expect to produce 2 million vehicles a year in Mexican plants. The emerging North American Free Trade Agreement (NAFTA) will move even more production jobs to Mexico. Perhaps the most devastating effect of job migration will be increased income inequality between the most and least educated workers. As Aaron Bernstein (1992, 48) wrote in *Business Week*:

> [A]s these people were laid off or suffered wage cuts, they created a glut of job candidates that helped hold down pay among the 64 million workers, across a wide spectrum of industries, who never went beyond high school.

It is difficult to assess the true impacts of NAFTA. Although most analysts agree that increased free trade will eventually bring more jobs to both the United States and Mexico in the short term, there will be dramatic alterations in the location of employment centers and wage rates in the United States.

The European Common Market as a trading system presents special challenges to the United States because the entire European market will be under a single market control. This will undoubtedly limit trade in some commodities and sectors. Ultimately, these trade agreements combined with the General Agreement on Trade and Tariffs (GATT) will have more profound impacts on workers than local economic conditions.

Capital Mobility

As mentioned above, capital is more mobile than people. Capital is a commodity and a producer service. International capital movements that manifest themselves in stock portfolio transactions, currency manipulation, firm takeovers, and buyouts are a major component of international/transitional corporatism. As Figure 1.4 reflects, the global restructuring of money leads to a disintegration of local economic independence and a marginalization of firms in many community environments.

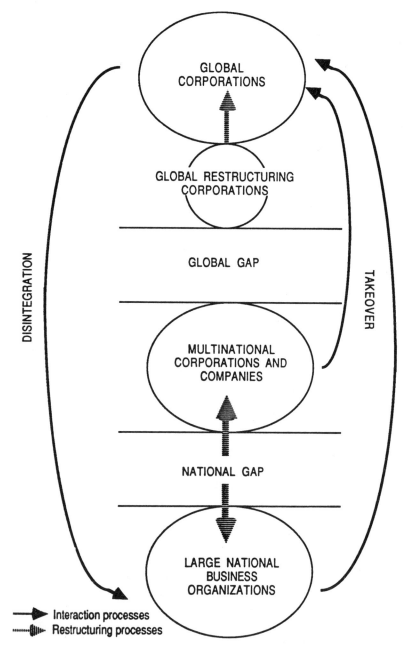

Figure 1.4. Organizational Segmentation and Corporate Interaction
NOTE: Circle size is approximately equal to the market significance of each group.

The driving reason for the use of capital for gain rather than the production of goods is relatively simple and fundamental. The profits are higher and the risk lower in portfolio dealing than in plant management. A systematic corporate strategy has emerged in the United States based on short-term profitability and increased control over the workplace. This orientation, according to some writers, leads to increasingly concentrated and centralized business, which has exacerbated unproductive administrative and supervisory costs, encouraged hierarchical and inflexible production methods, promoted shortsighted management decisions, and decreased worker job satisfaction and motivation (Bluestone 1984, 39). The deregulation of capital is one strategy to improve the available resources for local investment. The volatility of unfettered capital movements is reflected in the $500 billion savings and loan (S&L) disaster. The collapse of the S&Ls presents special challenges to the nation because it adds to the enormous debt and further constrains the nation's investment potential. The unregulated capital market strategy is meaningless, however, unless the following occurs: "More capital is likely to be optimal only if resources freed up are reemployed in activities of equal or greater productivity" (Bluestone 1984, 46). It is likely that freer capital will translate into more trading of currency and stocks rather than developing new products or creating new jobs.

In summary, the above factors represent a cross section of the forces causing communities to lose jobs and become increasingly insecure with respect to their economic future. These are not all of the factors or forces at work. Some other writers have combined these concepts differently and used different labels. Some analysts suggest several additional factors as being of equal or greater importance to those mentioned here. This analysis provides a sketch of the underlying issues associated with the current inability of the U.S. economy to produce the desired levels of wealth and consequent desirable employment growth. It provides a starting point for understanding the consequences of the current situation for communities and individuals.

The Consequences of the Changed Circumstances

Grim crime, family, and social dysfunctionality statistics reflect the consequences of the changed circumstances: rising discontent and hopelessness expressed by many U.S. workers and community activists. This helplessness and hopelessness is reported as increased alcoholism, family disintegration, homelessness, and incidents of crimes against property and persons. Joblessness leaves people feeling ashamed of themselves and their loss of status.

"Loss of a work network," as Bluestone and Harrison state, "removes an important source of human support. As a result, psychosomatic illnesses, anxiety, worry, tension, impaired interpersonal relations, an increased sense of powerlessness arise. . . . Unfortunately these tragic consequences are often overlooked when the costs of benefits of capital mobility are evaluated" (Bluestone and Harrison, in Staudohar and Brown 1987, 67).

Some people on the fringe of the economic system become permanently frozen out of the economic mainstream. The deindustrialization of the United States has created several new phenomena within U.S. communities in addition to these social costs (Root 1984).

A Segmented Labor Force

One of the consequences of these changes is the development of a more rigid, highly structured, and compartmentalized labor market. The emerging labor market contrasts sharply with a single labor market in which all laborers compete equally for available openings. The new national market, as Professor Bennett Harrison (1978) of MIT characterizes it, is highly segmented. At the core of the current labor market are the primary jobs. Primary employment consists of career professional and technical positions with good wages and benefits, career mobility, and additional training opportunities. Core jobs also tend to be unionized or have some degree of employment security. In addition, jobs in the primary sector are generally more rewarding than other positions because they tend to be "knowledge intensive." That is, individuals' intellectual skills are more likely to be used in the work.

Surrounding the core primary jobs are more marginal employment activities. The principal real employment in this segment lies in the so-called secondary labor market: low-wage, unstable employment in personal consumer services areas or in firms that are footloose. Much of this employment is both geographically and racially segregated. For example, Hispanics comprise the largest portion of the nondurable goods sector, African-American women dominate the lower wage service sectors, and Asians are moving into many personal services arenas. This work is not only segregated by sector but also spatially arranged within the metropolitan system (Mollenkopf and Castells 1991). Workers in these areas of the city are vulnerable to job shifts as economic cycles and lower-waged world labor markets produce volatile job movements. Furthermore, technological innovation might eliminate these jobs entirely. This form of employment segregation covers many occupational categories ranging from manufacturing to clerical areas. Individuals in this type of employment seldom have any job security and enjoy few benefits from their employers. Typical wages in these occupations average around $4.25 per hour, only marginally above the poverty line. As a consequence, individuals and/or families in most urban areas are forced to send more members of the family to work, including working-age children. (See Figure 1.5.)

Lying just outside the two employment areas are the employment staging areas of welfare and public sector job training. Often these employment segments work as a "revolving door." Individuals move from training into the lowest wage or the most insecure areas of the secondary labor market. Their tenure in employment is generally very short, which forces them back into training or returns them to welfare. Despite the enormous effort expended on such programs since the Nixon administration, fewer than 20% of the long-term unemployed persons entering training find unsubsidized nongovernment employment at the end of government-sponsored training.

The welfare segment of the labor market is a mixture of short- and long-termers. Individuals on welfare form a portion of the pool from which the training segment is drawn. Some welfare recipients use training to escape that system, and others become permanently mired in it. In 1990, there were 11.1 million parents receiving Aid for Families with Dependent Children (AFDC), the interim program developed by Franklin Roosevelt to assist families on a temporary basis. The long-term welfare recipient moves in and out of the other two legal employment areas. But wages in the secondary segment are usually too low to trade the security of welfare for the insecurity of a low-paying, frequently uninteresting job that people may have to travel a considerable distance to reach. Outside or on the fringe of the system are the *alegal* and the *illegal* segments. The alegal portion of the system is composed of individuals and even some small firms that do all their business in cash, pay few taxes, and drift along in the economic system. Occasionally, these cross over into illegal activities such as gambling or dealing in stolen property.

Illegal activities include a considerable underground employment market with its own rules and regulations. This market includes

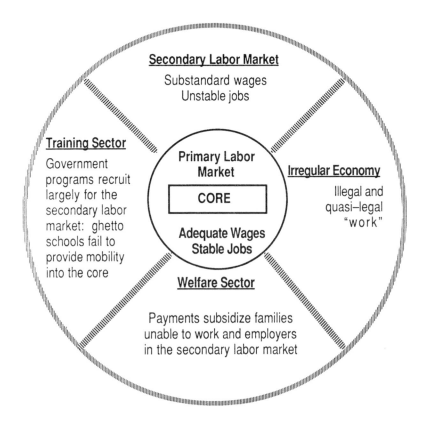

Figure 1.5. The Structure of Urban Labor Markets
SOURCE: Harrison, 1978. Used by permission.

drug dealing, prostitution, and various forms of small racketeering activities usually carried on at the neighborhood or community level.

Although individuals under the poverty line may drift into welfare, training, and occasionally secondary employment, many people in urban inner-city ghettos live a life of crime as their full-time regular employment. The risks associated with this lifestyle are well-known among its practitioners, and attempts at social reform have been especially aggravating.

The segmentation of the labor force, according to some, is becoming increasingly rigid. Studies by William Julius Wilson (Wilson 1987) and many others have graphically depicted this phenomenon. Fewer individuals have an opportunity to move into the regular core job market areas. Many Americans are permanently locked out of the economic mainstream because of the structure of the employment-generating system. The segmented labor market contrasts sharply with the so-called American Dream of upward mobility for all citizens.

Dual Labor System

The development of an information, service-based economy replacing the old industrial economy is a mixed blessing for communities and workers. For a small fraction of the emerging technology-based workforce, this new era means more and better jobs. But for many others, the jobs created in the information economy are worse. Considerable evidence suggests that the "middle is missing" in this employment system. That is, the economy now produces two distinct forms of employment. The 1979-1989 period (Figure 1.6a and 1.6b) is illustrative of this transformation: service employment in computing and similar areas, business service fields such as finance, real estate, and insurance, and retail trade lead the job growth pattern while heavily unionized sectors such as construction, manufacturing, and wholesale trade show modest growth or substantial declines.

Service employment is bifurcated into personal service, low wage jobs (hair care, hotel workers, etc.), and highly paid "good" jobs at the top for engineers and executives in the high-tech fields and their spinoffs. These jobs are "good" because the employee has considerable control over the product and processes related to the work activity. The most glamorous high-tech areas, it is estimated, will produce slightly less than 1 million new good jobs between 1990 and 2000. These workers will generally have intellectually stimulating employment and exercise considerable authority over their own work and that of others. They will also have greater general freedom to organize and control the flow of their labor.

At the other end of the system, the employment options and opportunities are less encouraging. The technology revolution stimulating the creation of new employment is also de-skilling and dehumanizing other jobs in both manufacturing and personal service employment. Even within the technology fields, the industry creates dull, uninteresting, and highly controlled work for the assembly-line worker. This same pattern is emerging in many other fields as well. Employees in clerical areas, retail sales, and manufacturing

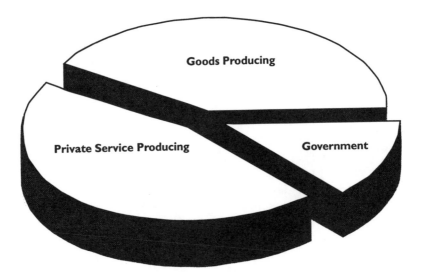

1950

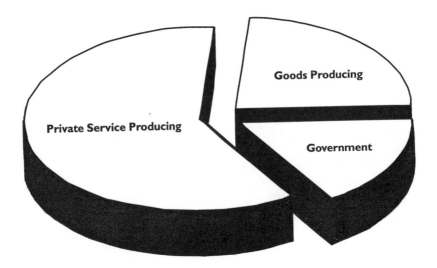

1989

Figure 1.6a. Distribution of Payroll Employment by Industry Sector, 1950 and 1989

SOURCE: *Monthly Labor Review*, September 1990.

Employment (in thousands)

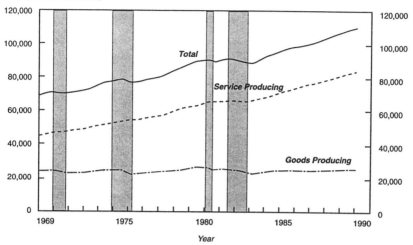

Figure 1.6b. Payroll Employment, Seasonally Adjusted, 1969-1989

SOURCE: *Monthly Labor Review,* September 1990.
NOTE: Shaded areas are recessionary periods, as designated by the National Bureau of Economic Research.

have far less control over the pace or character of their work. Because of technology, many positions are now interchangeable across the world because the computer-controlled equipment can be operated by almost anyone, anywhere.

As a result, the dual labor market is not only developing within the nation, it is taking on an international dimension as cheap labor and/or better machines replace current employees or at least reduce their control over the ultimate products. The jobs created for entry-level, lower skilled, and undereducated individuals are "bad" jobs in several respects. They are dull and require less worker input, which results in less job satisfaction. They are tenuous and subject to movement offshore or away from the local labor source. In addition, they are subject to technological change that can substitute machinery for people or processes. Finally, they are bad for the structure of the society because they offer little prospect for career improvement. Tenure and hard work in most of the new "bad jobs" simply do not place the worker in any better position to move up the company ladder because credentials and high-level knowledge-based skills are required for such advancement.

Thus, two classes or levels of employment are emerging. The career paths of the upper strata of the dual labor market do not intersect with the lower sections. This duality is not only across occupations, it also crosses international boundaries. Workers in certain advanced countries are now doing most of the thinking work while workers in less-advanced nations produce the final products.

The dual labor market may foreshadow the disappearance of the "middle" of the employment system. This could mean, some scholars argue, a reduction in the size of the nation's middle class with enormous implications for the socioeconomic order. There is increasing evidence that the middle class is shrinking. In fact, the Congressional Budget Office studies in 1990 indicate that the top 20% of the nation's families earned nearly 50% of all income while the bottom 60% of the population's income actually declined (Greenstein and Barancik 1990). Households earning a middle-class income fell from 28.7% in 1967 to 23% in 1983. This topic is the subject of considerable debate. What is indisputable is that both individuals and communities must make choices regarding economic development and employment planning. Communities have to determine whether new or expanding firms are improving the job base or merely reorganizing the existing workforce. Individuals have to ascertain whether every effort to create jobs will result in employment in a "real," lasting, "good" job.

Neighborhood and Community Decline

Inner-city urban neighborhoods and many rural areas have benefitted little from the rising affluence of the nation. Economic restructuring has very uneven spatial impacts both nationally as well as within communities. For example, California's San Jose, located in the heart of Silicon Valley, has pockets of persistent unemployment in its inner-city area. Inner-city neighborhoods have suffered long-term decline for nearly three decades. The reasons for the decline in these communities are complex. The consequences of the decline, however, are clearly visible in the form of high unemployment rates (particularly among nonwhite youth), crime, deteriorated shopping districts, and dilapidated housing stock. The root cause of this problem is systematic disinvestment in these areas over several decades. Several factors contribute to the loss of capital from these communities and the inability to attract new investment.

First, inner-city urban neighborhoods are victims of historically segregated housing and commercial patterns. Initially, factories, retail areas, and community residency existed in the same precinct. As zoning regulations based on the premise of segregated uses became more accepted, some factories were forced to move away from their labor base. This has made employment for inner-city groups more difficult as jobs move farther away. Almost simultaneously, white middle-income individuals began a corresponding move to the suburbs. This movement of affluent residents had its impacts on the retail base of the former industrial communities as nonwhites, principally African-Americans, moved into the communities. The loss of firms and white residents caused banks to engage in a practice of "redlining" the zone for future investment or of actively divesting themselves of holdings in these neighborhoods. Capital flight and disinvestment became the dominant pattern in economically unhealthy inner-city areas. The lack of public investment in the same communities soon manifested itself in the deteriorating quality of local services, housing, schools, and community facilities.

Second, as inner-city neighborhoods lost their retail base, they also lost employment for their residents. As a result, poor communities were poor markets. The loss of market functions in the neighborhoods reinforced the general malaise of lowered community pride. The absence of local markets meant not only lost services but also lost income. This loss of local and external capital compounds the localities' problems, because employment opportunities require investment dollars.

One counterinvestment trend in inner-city areas has been the investment in social welfare facilities ranging from housing and social service agencies to community-based criminal parole centers. The relocation of these facilities to the inner city may have been well-intended, but the result has been disastrous. As social services and welfare services move into a community, many established residents have elected to move to the suburbs or more affluent areas of the city, leaving behind more people who require public social services. In essence, these inner-city/minority communities became socioeconomically skewed toward dependence with a consequential loss of the internal capacity to develop themselves.

Third, the federal government itself has disinvested in urban areas with severe consequences to inner-city neighborhoods. In the 1970s, the federal government poured nearly $60 billion a year into urban aid projects ranging from community revitalization projects to schools and job training. The federal government spent over $5 billion on public service job training alone. These funds almost disappeared in the 1980s and as a result the hope and spirit associated with urban recovery seemed to fade as well.

Finally, in the 1980s, inner-city residents became viewed as lacking initiative, skills, and motivation to work and lacking the social capacity to fit the modern workplace. Racial and sex discrimination remain a component of this problem. The urban ghetto resident, however, must also overcome the stigma of his or her address. Discrimination by address is a new and widespread practice among many employers. The inner-city areas are often viewed as places of criminality, low motivation, and poor education. Employers in tight job markets use assumptions about where people reside as an additional screening device.

The Rural/Small Town Dimension

Rural and small town poverty and unemployment are as severe as urban poverty and unemployment. By some measures, they are worse. Despite the tremendous growth of nonmetropolitan employment mentioned earlier in this chapter, many rural areas and small towns have been hard hit by the slowdown in both manufacturing and agriculture. In fact, nearly 500 rural counties in the 1990 census were areas of continued and persistent poverty.

High interest rates, farm land speculation, and overproduction have severely overtaxed many rural community institutions, including the local financial base. As farms collapse, mines close, and other mineral extraction operations diminish, the dwindling employment base in agriculture and extractive areas declines even more swiftly. The loss in farm, timber, and mining income has a ripple effect through an entire community.

In addition to declining demand for agricultural, oil, and other natural resource products, many rural communities have experienced manufacturing branch plant moves and/or closures. Branch plant closings have had a more marked impact on rural than urban communities. In fact, many single-crop and single-firm nonmetropolitan communities are extremely vulnerable because of their narrow economic base. Many predominantly rural Southern and Western states with large component manufacturing or assembly plants suffered even higher rates of plant closings and overseas movement than urban areas.

Small towns and rural areas have suffered from the twin impacts of the manufacturing decline and the farm crisis. As branch plant economies, these communities exercise little control over the corporate decision-making process. When international commodity prices fluctuated, the actions of other suppliers of raw materials influenced the base economy. In a sense, some rural communities have been trapped by their own success in attracting firms and expanding foreign trade in agricultural and mining commodities. Irrespective of the origins of these problems, small towns must adjust. Tourism and retirement will not replace the jobs lost in the basic sectors of rural economies. Although efforts to revitalize rural areas by making small towns cute or quaint are appealing, in the last analysis, much more is required to strengthen the economic viability and job potential of the nation's heartland.

The Rising Underclass

A new subculture is rising in the United States: the *underclass*. This growing subgroup is composed of individuals trapped in a cycle of poverty, unemployment, unstable families, and petty criminality. Members of the group are predominantly nonwhite and intergenerational welfare recipients.

Precise definitions and the size of the underclass vary according to data source. Although the dimensions of this group are uncertain, there is no disputing its existence. In 1990, over 33 million Americans were living below the official national poverty line of $13,924 for a family of four. Of these, slightly more than 17 million were non-Hispanic white families. Despite popular stereotypes, most poor people are white—not black, brown, or yellow. Poverty and consequent underclass problems are not a race issue. Persistent white poverty is increasing at an alarming rate in both metropolitan and nonmetropolitan areas. Unfortunately, the nation's cities are becoming areas of concentrated poverty and low opportunity, as depicted in Table 1.1.

According to welfare experts, about 6.7% of the poor are mired in long-term irreversible poverty lifestyles (Jencks and Peterson 1991, 37). This group has a disproportionate share of urban African-Americans and Hispanics within it who are divorced from

Table 1.1   Indexes of Residential Segregation for Selected Metropolitan
Areas with Large Minority Population, 1990

Metropolitan Area	% Minority Population	Index of Segregation of Non-Hispanic Whites with:		
		Blacks	Hispanics	Asians
Northern and Western Areas				
Boston	15	72	59	46
Buffalo	15	84	60	56
Chicago	38	87	65	47
Cincinnati	15	79	36	47
Cleveland	23	86	57	42
Columbus	15	71	34	49
Detroit	25	89	42	48
Gary-Hammond	28	91	53	42
Indianapolis	16	78	32	43
Kansas City	17	75	42	39
Los Angeles-Long Beach	59	74	63	48
Milwaukee	19	84	58	47
New York	52	83	68	52
Newark	36	84	67	35
Philadelphia	25	81	65	47
St. Louis	20	80	29	44
San Francisco	42	66	51	51
Southern Areas				
Atlanta	30	71	39	45
Baltimore	29	75	35	42
Birmingham	28	77	37	53
Dallas	33	68	54	47
Greensboro-Winston Salem	21	66	35	49
Houston	44	71	53	50
Memphis	42	75	41	43
Miami	70	72	52	31
New Orleans	41	72	34	54
Norfolk-Virginia Beach	33	55	33	37
Tampa-St. Petersburg	17	74	47	39
Washington	37	67	43	35

SOURCE: Data from "By the Numbers" 1991, 3a.
NOTE: Dissimilarity Index: 100 = complete segregation, with no mixing of races in same census
tract.

the community aspirations associated with family, employment, careers, and the promise of education.

The saddest and most disturbing feature of the underclass is the teenage pregnancy rate among its members. It is not uncommon for women of 32 to become grandmothers as their unwed daughters, having barely entered their teens, have children of their own. Teenaged unwed mothers account for almost all of the 1 million welfare recipients receiving child support (AFDC).

Teenage pregnancy leads to high dropout rates in the schools, reinforces family breakdown, and prevents upward economic mobility. Yet the problem is endemic in the African-American community, where the majority of children are born into fatherless homes to underaged and underprepared parents. The lives these newborns begin are not especially optimistic. Many are underweight at birth and suffer from assorted physical and mental disabilities. They have no role models in the traditional family or work world. They are unlikely to complete formal schooling.

Females form the bulk of the new underclass for fairly obvious reasons. Female-headed households constitute over 46.5% of the persons under the poverty line. These women have low skills; if and when they work, their incomes still remain low enough to qualify them for public assistance. *The feminization of poverty* is not a new or startling finding. Female poverty has been the most striking feature of the social welfare system almost since its inception. What is new is the persistence of this poverty among certain groups of women, predominantly African-American and Hispanic women, who cannot find a way out for themselves or for their offspring.

Undereducation as well as illiteracy are also central features of the underclass. A national assessment of literacy indicated that 12.6% of all Americans over 17 years old were functionally illiterate. Of this group, it is estimated by various sources that 41.6% of African-American youth under 17 are functionally illiterate and 82.7% only semiliterate. These undereducated youth fall predominantly in the welfare/dependent groups described as the core of the underclass. In a computer-advanced technological era, individuals without adequate reading and related skills cannot compete for legitimate work. Efforts to incorporate them into the work world by merely increasing the number of jobs available are doomed to fail.

Firm Migration and Reorganization

Local businesses priding themselves on their community conscientiousness have reassessed this position in light of their circumstances. A firm cannot pledge its loyalty to any community or workforce if taking that position means failing to remain competitive and gain a return for the stockholders. Firms not only must reassess their commitment to location but also assess their position with the local labor force. Even attempts at corporate good citizenship must be tempered by the prospect of hostile takeover bids and international competition. No company is safe. As a result, even good firms, like some unemployed people, have become transitory migrants in the community. Some companies have been forced to shift their corporate headquarters from their national base city. This was unthinkable only a decade ago, but it is common today.

Corporations have become nomadic. Literally thousands of corporate headquarters have moved away from their home cities and hence far away from their employee base. Even *Fortune 500* firms have elected to move away from major cities such as New York and Los Angeles to suburban locations or even overseas. Some moves to the suburbs or growing metropolitan areas were to secure a better labor force or a greater share in growth area markets; but many more were the result of external factors related to corporate takeovers and similar activities.

Firms are not only moving away but also reorganizing away from their current workforces. For example, steel companies have expanded into the oil or liquor business, and steamship companies operate hotels. Some new company forms are so complex that the firm's precise line of business is not clear, such as TransAmerica Corporation or ITT. These firms shed and add workers to meet their current requirements rather than retraining hired workers.

Clearly, the existing structures within communities—and even within nations—for coping with the rapid transformation of the international corporation are inadequate. Nonetheless, every community is faced with the prospect of being only a momentary address for a major employer. When the firms move, the people stay behind. Communities must build economies that are competitive even in this environment.

Summary and Conclusion

What does all this mean? Are the circumstances so difficult and overwhelming that local decision makers surrender the community's fate to chance or the marketplace? Can anything meaningful be done to incorporate the unemployed and underemployed into the employment system? Are there enough jobs? Are there enough "good" jobs? Can a single community tackle the problems of joblessness, homelessness, or declining job base by itself?

There are, of course, no clear answers to these questions. What is clear is that communities must have a better perspective on the problems affecting the national and international economy and their consequences for the citizenry in general and for certain groups in particular. Local economic development and employment generation can be designed in any locale to deal more effectively with these problems. Understanding the dimensions of the national phenomenon and its consequences for a locality are the first steps in organizing realistic courses of action.

The central thesis of this book is that locally based economic development and employment generation is more likely to be successful if initiated at the community/local level rather than elsewhere. Each of the factors influencing the economy has unique manifestations and slightly different causes in each local area. General solutions to community problems will not succeed if they are not targeted to specific groups and linked to the total regional economic system.

Community leaders can identify the situation their area faces and place it within a larger context. Similarly, an assessment of the groups affected can be made in order to determine how various groups will respond to different courses of action. In this context, local solutions can be found for national problems. The next chapter explores the dimensions of national policy that currently serve as a basis for local economic development.

References and Suggested Reading

Auletta, Ken. 1982. *The Underclass*. New York: Vintage.

Bergman, Edward, and H. A. Goldstein. 1986. Dynamics of Structural Change in Metropolitan Economies. *Journal of the American Planning Association* (Summer).

Bernstein, Aaron. 1992. The Global Economy. *Business Week*, August 10.

Blakely, Edward J., and Ted K. Bradshaw. 1986. Rural America: The Community Development Frontier. *Research in Rural Sociology and Development* 2:3.

Blakely, Edward J., and Philip Shapira. 1984. Industrial Restructuring: Public Policies for Investment in Advanced Industrial Society. *Annals of the American Academy of Political and Social Sciences* 475:96.

Bluestone, Barry. 1984. Is Deindustrialization a Myth? Capital Versus Absorptive Capacity in the U.S. Economy. *Annals of the American Academy of Political and Social Sciences* 475:39.

Bluestone, Barry, and Bennett Harrison. 1982. *The Deindustrialization of America*. New York: Basic Books.

By the Numbers: Tracking Segregation in 219 Metro Areas. 1991. *USA Today,* November 11, p. 3a.

C&R Associates. 1977. *Community Costs.* Washington, DC: Author.

Cohen, S. S., and J. Zysman. 1987. *Manufacturing Matters.* New York: Basic Books.

Goldsmith, William, and Edward J. Blakely. 1992. *Separate Societies: Poverty and Inequality in U.S. Cities.* Philadelphia, PA: Temple University Press.

Greenstein, Robert, and Scott Barancik. 1990. *Drifting Apart: New Findings on Growing Income Disparities Between the Rich, the Poor, and the Middle Class.* Washington, DC: Center on Budget and Policy Priorities.

Harrington, Michael. 1963. *The Other America: Poverty in the United States.* Baltimore: Penguin.

Harrison, Bennett. 1978. *Job! What Kind, for Whom and Where?* Cambridge, MA: Massachusetts Institute of Technology.

Jencks, Christopher, and Paul Peterson. 1991. *The Urban Underclass.* Washington, DC: The Brookings Institution.

Litvak, Lawrence, and Belden Daniels. 1979. *Innovations in Development Finance.* Washington, DC: Council of Planning Agencies.

Metropolitan Applied Research Center. 1968. *A Relevant War Against Poverty.* New York: Metropolitan Research Council.

Mollenkopf, John, and Manuel Castells. 1991. *Dual City: Restructuring New York.* New York: The Russell Sage Foundation.

Noyelle, T. 1983. The Rise of the Advanced Services. *Journal of the American Planning Association* 49(3).

Organization for Economic Cooperation and Development (OECD). 1986. *The Revitalization of Urban Economies.* Paris: Author.

Pigg, Kenneth. 1991. *The Future of Rural America: Anticipating Policies for Constructive Change.* Boulder, CO: Westview.

Redburn, S. F., T. Buss, and L. Ledebur, eds. 1986. *Revitalizing the U.S. Economy.* New York: Praeger.

Reich, Robert B. 1991. *The Work of Nations: Preparing Ourselves for the 21st Century.* New York: Knopf.

Root, Kenneth. 1984. Human Response to Plant Closures. *Annals of the American Academy of Political and Social Sciences* 475:52.

Staudohar, P., and H. Brown. 1987. *Deindustrialization and Plant Closure.* Lexington, MA: Lexington.

Sternleib, G., and J. Highes. 1975. *Post Industrial America: Metropolitan Decline and Interregional Job Shifts.* New Brunswick, NJ: Center for Urban Policy Research.

Strange, Walter. 1977. *Job Loss: A Psychosocial Study of Worker Reaction to Plant Closing in a Company Town in Southern Appalachia.* Baltimore, MD: National Technical Information Service (NTIS).

Summers, G. F., ed. 1984. Deindustrialization: Restructuring the Economy. *Annals of the American Academy of Political and Social Sciences* 475 (September, entire volume).

U.S. Department of Commerce. 1977. *The Development of a Sub-National Economic Development Policy.* Washington, DC: Author.

Wilson, William Julius. 1987. *The Truly Disadvantaged: The Inner City, the Underclass and Public Policy.* Chicago: University of Chicago Press.

Varaiya, Parvin, and Michael Wiseman. 1981. Investment and Employment in Manufacturing in U.S. Metropolitan Areas 1960-1967. *Regional Science and Urban Economics* 11:431.

2 | National Policy Options and Local Economic Development

Imperfect Solutions to Local Problems

No community is an economic island. All cities and towns, rural and urban alike, share the same economic plight. Until 1974, as discussed earlier, rural and urban areas were much less internationally dependent. In fact, American communities could rely exclusively on their regional and national market positions to determine their economic stability. This is no longer the case. Cities, no matter how large or small, are now part of a global economic system. In this system, regions within nations form the economic building blocks. As a result, cities, suburbs, and rural areas have become linked into a world economy. In many instances, such as in the cases of Los Angeles, New York, San Francisco, and Miami, the links these regions have to the international economy are more significant than their ties to the domestic economy. In essence, the factors influencing local economies are international in dimension, with varying ramifications for each community and for the economic fabric of the entire nation. In these circumstances, Peter Drucker warns:

> Practitioners, whether in government or in business, cannot wait until there is new theory. They have to act. And their actions will be more likely to succeed the more they are based on the new realities of a changed world economy. (Drucker 1991, 4)

Drucker outlines the factors that are *uncoupling* national and local economies from one another. This uncoupling has propelled metropolitan areas into international competition with one another, using

their own resources. Moreover, he points out that the most fundamental shift in the world economy has been the shift from labor-
intensive to knowledge-intensive industries. These changes in the
world economy are increasingly reflected in cities and towns all
over the nation.

Since the Clinton administration took office it has had to grapple
with increasing international dangers to the domestic economy.
The Japanese real estate bubble burst in 1992, sending shock waves
throughout the Western industrialized nations. The national debt
threatens the country's ability to invest in productive enterprise.
In 1992, the debt stood at $4 trillion. Interest payment on the debt
is now the second largest federal expenditure at $292 billion or
$2,299 per worker. This kind of toll on the national capacity cannot
be ignored. Some believe that the nation must tighten its belt by
reducing expenditures, especially entitlements such as health and
social security, along with curtailing military and related domestic
spending. In addition, President Clinton proposed a new jobs bill
in 1993 to help the nation produce its way out of the debt by
investing in the infrastructure required to become competitive in
the next century. This bill failed in Congress. No matter what
course(s) of action are taken there is little doubt that sacrifice will
be required. Local communities will have to do their part to stretch
current resources and find ways to increase public and private
productivity.

In order to restore competitiveness, the Clinton administration
Jobs Program proposed policies to reduce unemployment and
retrain workers for new enterprises. These policies must deal with
the reality that led the U.S. Bureau of Labor Statistics in 1988 to
estimate that it would take a growth rate of over 5% annually,
without significant inflation, to return to pre-1974 employment
levels. Furthermore, the *jobs crisis* has a clear geography. Its spatial
dimension—depressed inner cities and abandoned factories in the
largest metropolitan areas, contrasted with rising employment
and affluence in the suburban ring—causes policymakers the most
difficulty. Current attempts of national policy tools are demonstrat-
ing the futility of macro-economic solutions in confronting local
problems. Even when local supportive efforts, such as local policy
boards, are added to a national program, the programs seldom resolve
the underlying cause of regional or local economic and employment
problems.

There is little argument that the national government has a
responsibility to act to stimulate the economy in order to stabilize

and improve both national and local employment. Although national economic policy is important, however, it has severe limitations with regard to meeting the twin needs of economic sectoral adjustments and regional/local employment requirements. Further, few effective national policy tools will assist in basic jobs formation. The major economic stimulators the federal government currently uses are designed to facilitate overall economic recovery rather than to address specific segments of the population or localities. The policy strategies that the federal government uses to respond to the current job crisis are reviewed in this chapter to show how limited they are in responding to the problems at hand. For this reason, the Clinton administration is embarking on a more regional and local economic development approach, including such measures as regional economic policies, neighborhood revitalization, and community development banks.

The National Industrial Policy Debate: "Adhocracy" in Action

In the wake of the national economy's continuing malaise, the Clinton administration has adopted a set of industrial policies to stimulate the national economy. Unlike recent national administrations, the Clinton Cabinet has embraced national industrial and social policies as cornerstones of economic recovery. Such proposals are scarcely novel. The Louisiana Purchase, the development of the railroads, the creation of the land grant college system, and the Depression-era New Deal were all industrial policies. They were pursued with vigor and used whatever national resources necessary to obtain the desired economic results. Many of these programs, like the New Deal, were controversial when they were introduced. Nevertheless, stimulating the national economy using federal government tax, financial, regulatory, and monetary policy is as "American as apple pie." Yet economists still debate whether national policy should direct or even try to stimulate a market economy. Even more skepticism exists over the effectiveness of a national industrial policy in targeting particular disadvantaged groups or in dealing with the spatial dimensions of unemployment such as in inner cities, the Rustbelt, and defense-industry-dependent communities. Nevertheless, failure to incorporate these areas and groups into the national economy would clearly have severe repercussions on the entire nation for many years to come. Yet today, the efficacy of national industrial policy in dealing with the issues described has spawned a heated policy debate.

On one hand, some economic reformers invoke past traditions with regard to economic planning. A contingent of these reformers favors nationwide reindustrialization. They want to rebuild the nation's industrial stock through new labor regulations, a new set of targeted tax incentives, and nationally financed infrastructure development. This group would refocus economic development programs on rebuilding the nation's industrial plant in the sectors where the nation must remain internationally competitive (Cohen and Zysman 1987).

One of the chief proponents of this point of view, Secretary of Labor Robert Reich, has proposed reinvestment in the American people as the basis for national recovery. He believes that the transformation from old to new industrial and commercial activities can be accelerated through appropriate investments in infrastructure and tax incentives for new high technology businesses. Reich and his supporters would identify the "sunrise" industries as soon as possible and provide them with the human resources to succeed. The principal prescription for such industries requires, in their view, the dismantling of the current ad hoc approach to economic development planning for a more coherent, targeted, and planned approach. As the Clinton administration's tax plan proposes, tax concessions and other direct and indirect investments might then be funneled to the emerging firms that produce new jobs that compete in a global market. Clearly, some form of management system needs to be put in place to guide this system from the national level, as Reich put it so well:

> Stop fighting over how much money government is taking from the wealthy and redistributing to everyone else. Start worrying about the capacity of Americans to add value. (Reich 1991, 52)

One version of the Reich approach is the *reconstruction finance corporation* suggested by Felix Rohatyn and others. This agency would provide the guidance system for mobilizing national resources and picking the winners in economic development. An even more corporatist strategy has been proposed by Ira Magaziner, who suggests an explicit compact between business, labor, and government to guide national as well as local development. At the state and local levels, this form of corporate economic strategy planning has been endorsed by many officials. The U.S. Conference of Mayors praised this approach because it tends to increase local participation in economic recovery.

An opposing point of view led by a number of prominent public finance scholars take a dim view of incentives and dedicated taxes. They favor less—rather than more—government involvement in economic development and industry. The *less government* advocates warn the more interventionist groups of the follies of their past actions, which have essentially made the nation a victim of regulations, of special support systems, and of inconsistent, often contradictory, industrial and commercial policies that restrict trade and make the nation less competitive. Instead, the group supports a move back to a fundamentally free market. In the American market economy, the best can and will survive and the nation will be better served. Some adherents of this view would like to see the labor market deregulated even further as a means of absorbing the unemployed groups in the nation. In essence, if wages found their natural level, a lot more jobs would be available for those genuinely willing and able to work.

In contrast to both the views of the free or planned market approaches, some planners and economists impressed with European economic systems are proposing industrial policy that deals explicitly with localities and people rather than firms and capital. This group suggests that free market or regulated industrial policy perspectives are misconceived because they are based on the false premise that "what is good for business is good for communities and workers." Although there is some merit in assisting in firm expansion, an alternative view advanced by myself and others is that industrial policy must have a spatial dimension as well. Specifically, our argument is that (1) national policies are needed to increase community control over corporate investment policies, (2) communities should have a greater role in determining their economic stability and quality of life, and (3) workers should have increased control and certainty over their livelihoods.

The means to reaching these ends are not particularly radical. They include national policies that restrict tax write-offs for nonproductive investments; that provide government assistance in community economic development planning, including loans to achieve both ownership and development of industrial space at the community level; and that provide worker retraining credits and/or educational incentives as well as more portable retirement and benefits. In essence, each worker should have an employer-supported transportable retirement account outside of

Social Security, to which any employer could contribute and to which the worker could also make contributions. Other operational policies are also suggested (Blakely and Shapira 1984).

Although many economists deny it, the nation has a crazy quilt of de facto or ad hoc industrial policies. The substance of these policies ranges from farm supports to tax advantages for certain investments. States and cities have added their own taxes, land subsidies, and similar incentives to these national efforts. The net effect of the existing approach is bewildering and sometimes counterproductive. Nevertheless, the policies represent the current policy/nonpolicy approach to national economic recovery. Communities all over the nation are developing their own industrial policies using various approaches. Industrial policy at the national level means little unless there is companion policy at the local level to take advantage of available federal and state resources. At long last, a specific national industrial policy is being enunciated. The principles of the new national policy incorporate many of the methods described in this book.

Monetary and Tax Policy

In the 1970s the Federal Reserve Board began promoting a set of economic management approaches based on controlling interest rates. These policies have led to a relatively high cost of borrowing for the nation's business. They are designed to combat inflation and deal with a high-consumption/low-production economy, which has succeeded for some and failed for others. For new small businesses, these policies have meant money was tight. For others, the reduction in the inflation rate and the economic stability have increased business confidence and stabilized both wages and prices. Federal policies have achieved the basic mission of controlling rapid economic growth. But this has neither led to strong job growth nor arrested plant closings or inner-city neighborhood disinvestment. In large measure, the problems associated with the demise of the nation's industrial economic character and weak competitive position cannot be overcome by improving fiscal policy alone.

Demand for U.S.-made goods remains high. The United States has expanded its export base. In 1991, the country exported over $422 billion of industrial and agricultural products and over $164 billion in services, making the U.S. the world's largest exporter. Nonetheless, the United States has one of the world's largest trade deficits.

The trade deficit is over $60 billion with Japan alone. Emerging industrialized nations such as Taiwan enjoy the world's largest foreign exchange. This island nation holds over $150 billion (1992) based almost entirely on its huge U.S. trade gap. The reason for this is simply that foreign goods remain relatively cheap and superior in quality to domestic counterparts. Monetary policy alone will not address this problem. U.S. management must become sensitive to the customers' demands both here and abroad in order to compete successfully.

Tax policy has always been a major component of economic development policy. Federal taxes are sophisticated instruments designed to steer private investment capital. Tax write-offs and loopholes are the primary motivators for short-term investments. Some economists argue that the nation could use tax policy exclusively to redirect much of the nation's wealth into productive enterprises. Foreign investments are also a potential for pretax dollars. Europeans invested $35 billion in the United States in the first 9 months of 1989. By 1992, this investment flow had slowed to under $11 billion a year. Therefore, reindustrialization may be either close at hand or distant depending on where the new capital seeks a return. The Clinton administration wants to find better ways to tax foreigners so they pay their fair share of the nation's economic recovery. These tax policies combined with a slow market may influence capital flows to the United States.

Whatever direction capital moves, communities must help businesses by planning carefully and aggressively to assist new firms, to rebuild old ones, and to meet the needs of firms seeking new markets or developing new products. In other words, capital has to be captured—it will not seek investment.

Welfare and Social Policy

The literature on social welfare has generally been critical of both the goal and the result. Welfare policy has been criticized for having created the underclass described earlier. Some evidence supports this contention. In 1961, over 65% of African-Americans below the poverty line were engaged in some form of work. Moreover, most were in intact family situations despite their poverty. By 1971, only 31% of African-Americans below the poverty line were engaged in any form of legitimate employment. Most of the African-Americans below the poverty line were in irregular family situations and rapidly drifting toward underclass status. By 1985, the total central-city population living in poverty reached 20.9 million, almost double the 1970 poverty population (Goldsmith and Blakely 1992).

National welfare policy was always intended to be temporary. Its evolution into a permanent feature of American life has astounded and confused policymakers and social workers. Welfare recipients are not moving into employment. In fact, nearly one third of the nation's welfare recipients are from welfare homes. The general public, including many poor people, hold welfare in contempt and view recipients as principally lazy and licentious persons. Loss of public sympathy combined with a shrinking employment structure has led national and state policymakers to experiment with welfare, combined with education and work, as a path to employment. A number of schemes have been put into effect. Their central theme is to introduce the welfare person into the work world and to correct bad work habits or prevent them from developing.

Workfare, as this approach is called, aims at making welfare recipients increase their employability skills in exchange for their dole payment. Another version of this approach is community-serving enterprises that employ welfare recipients as the workforce. In this approach, a nonprofit agency becomes the employer of the welfare-eligible person. The employer receives the welfare payment plus a premium to provide social services and in turn places the welfare person on the payroll of a regular but highly subsidized job.

The results of both of these approaches have been mixed. Nevertheless, they compose an alternative to being locked into the underclass. As economic development policy tools, they offer the potential to raise the human capital among the urban area population, thereby improving the possibility that those areas might attract or retain an economic base.

Employment and Training Policy

The most innovative area of economic development policy in recent years has been in the field of employment and training. The introduction of the Comprehensive Employment and Training Act (CETA) of the Nixon administration initiated a new wave of employment policies driven by economic development.

Earlier employment and training plans were designed to improve the employability of the "hard core" unemployed or to improve the ability of certain areas of the country or segments of the population to enter the job market. The problem was perceived as a deficiency of skills in an abundant job market. Training would and could solve the skill dilemma. This was only partially true at the time the program was conceived. Discrimination, lack of social skills, and poor aggregate job formation in lagging regions of the country or inner-city areas prevented most people from entering the job market. Nevertheless, the investments in human capital formation were appropriate for the time.

The problem today and for the foreseeable future is that there are not enough jobs and clearly not enough "good" jobs. As a result, training does not help. Therefore, employment planners have become concerned with increasing the number of jobs and improving the regional economy to absorb the existing labor force. The CETA program and its successor, the Jobs Training Partnership Act (JTPA), build expressly on the notion that economic development and employment must be matched.

The emphasis of the Clinton administration has altered this approach to one that attempts to build human capital into new enterprise formation: in essence, building jobs around the existing population rather than trying to import jobs to them. The administration hopes to achieve this objective by combining enterprise zones with welfare reforms, inner-city infrastructure, and school improvements.

This change in emphasis is perhaps the first attempt to move away from the old training model that relied on the expansion of large employers. Earlier overt economic development programs to assist neighborhoods, inner cities, and rural areas with physical improvement or industrial support were drastically reduced or simply eliminated in the Reagan-Bush era. The Clinton administration has embarked on a massive effort aimed at rebuilding communities and the local skill base. Thus, the federal government is altering its posture toward job formation and human capital development. The new policies aim to build jobs where the people are and to increase their skill so that they can compete with foreign labor.

Trade Policy

The United States is experiencing the worst balance of payments (incoming goods versus outgoing goods) in the history of the nation. At one time in the recent past, the demand for U.S. products was so great that we could scarcely keep pace with it. At that time, we were champions of "free trade." The general belief was that the best can and will maintain their markets. Tariff protection served only to protect weak industries and underpinned weak economies. Now the shoe is on the other foot. U.S. goods are not penetrating foreign markets. To some extent, the trade policies of the European Common Market, Japan, and other major trading partners are not as open as those of the U.S. market. The North American Free Trade Agreement may provide a new platform for the United States in international trade competition. The U.S. dollar, as the world currency, has been maintained at such artificially high levels that it has been difficult for U.S. manufacturers to compete in overseas marketplaces. But to a great extent, production in other countries is catching up with internal domestic demand and quality is improving, although living standards remain modest in these nations.

U.S. labor unions, farmers, and even some businesses are calling for the protection of American firms from international competition. In many instances, such as with Japan, this is a retaliatory move against the other country's own import quotas and duties on U.S. goods. In other cases, the desire for higher tariffs is a response from an inferior competitive position. Unions and some communities view high tariffs and import quotas as the only mechanism for saving jobs. These approaches will not save jobs in the long run. Although this approach may be shortsighted, it is an important current remedy for these workers and their firms, as well as for the members of Congress who represent the communities. As a result, new tariffs and trade regulations are often promoted as a major component of industrial policy.

A companion to the creation of shelters for domestic industry is the creation of export subsidies for agriculture. The rationale for this is that U.S. agriculture is in competition with more heavily subsidized farmers in Europe, Australia, and Asia for world markets. The GATT agreement on agriculture was almost killed by very highly subsidized French farmers. Whether or not the subsidy arguments have any economic rationale, they are politically volatile. Congress has yielded to protectionist sentiments by creating large export subsidies for agriculture.

A substantial body of recent research indicates that the most heavily protected manufacturing firms have been the least competitive internationally. Clearly, the use of tariff protection and subsidies will mean some short-term improvement in various communities. As the literature on trade policy and production suggests, however:

> It is well established in economic literature that tariffs or quotas are not the best policies to achieve goals related to national or sectoral production and employment . . . [and] are even less preferable for achieving regional employment goals. (Ohlsson 1984, 14)

For these reasons, some economists contend that protection-oriented trade policies cannot discriminate sufficiently to protect a nation's competitive enterprises from less productive firms still in existence as a result of tariffs. Furthermore, tariff protection as a national economic policy merely locks dying, uncompetitive firms and industries into some regions, which only undermines long-term economic viability by discouraging community-initiated diversification into more competitive sectors of the national/international economy.

Regional/Local Development Policy

Development policies for lagging areas were the cornerstone of the New Deal. The development of the Tennessee Valley Authority, Rural Electrification, Cooperative Extension, and a plethora of programs aimed at rural underdeveloped areas have become part of the pattern of American life. In the 1960s, Presidents Kennedy and Johnson merely extended these good ideas to the inner cities in order to stimulate internal development of urban neighborhoods and central business districts hard hit by the development of the suburbs. These programs have been extended and modified over the last several decades, but their key features remain intact.

In general, almost all federal *development* efforts have had a "bricks and mortar" orientation. The role of the federal government has been to provide the physical conditions and infrastructure to induce development rather than to provide direct intervention in the private sector. Some marginal interventions have been companions to these physical programs, such as the location of military facilities and government offices. Federal efforts to stimulate economic development among the poorest regions and groups to date have been indirect (except for the War on Poverty) in their application to community problems.

Both the federal government and local officials have tacitly agreed that national government should remain in the background, using its money but not its muscle to bring about economic change. The mixture of funding and intervention has become more difficult with the decline of federal funds available for local use. As the nation's industrial stock diminishes, local officials have more need for federal assistance; however, less assistance is forthcoming. Military base closures will affect over 150 communities by the turn of the century. Local and state governments are hard-pressed both to meet social commitments to the unemployed and to stimulate develop-ment. The old inducements of tax holidays and the like have to be balanced against a shrinking (rather than expanding) revenue base.

The federal government's response to this problem has been to reduce the red tape and strings attached to the remaining funding. Essentially, the national government has given local officials more authority to deal with fewer resources. This gives many local residents the impression that local policymakers have more discretion over federal funds and more options to cure local problems. The simple truth is that there are more problems than money. Local officials have to use the funds they have as investments in the future, not as temporary aids to meet current emergencies. Herein comes the need for, and the difficulty in, engaging in local economic development.

The federal government's response to the needs of communities and individuals is under significant revision. The new policy struc-ture is still emerging. It is too early to tell whether this set of policies will have the desired impact. However, communities can-not afford to risk a laissez-faire approach to their own destiny. Local policymakers must take constructive actions to meet residents' needs in both the short and the long run. National policy may be well-suited to the long run. Communities faced with plant closures, high unemployment, rising crime, family disintegration, and increas-ing public assistance burdens must face the short run by taking steps to deal with their economic destiny sooner rather than later.

National Economic Development Policy

As the earlier discussion suggests, national policies with real employment consequences must improve the ability of firms to compete and increase the capacity of communities to build employ-ment. National government can fashion policies that enhance the capacity of regions and localities to pursue development strategies that will strengthen their own competitive advantages as well as those of the national economy. As Carter (1984, 7) suggests:

> [National] industrial policy will be more effective if coupled with active regional [and/or local] policies centering on the local implementation of programs of productivity improvement and support measures that assist regions to develop economic activities that utilize their comparative advantage.

Policy measures may be classified according to their implicit or explicit discrimination between regions, sectors, or firms, as shown in Table 2.1. The policy alternatives depicted represent the ways that policymakers, both in the United States and elsewhere, think about the place and means of intervening in regional/local economic policy. This country has tried variations of all these approaches during different periods. It would be safe to say that the situation governs the response in the United States. There is no one way to do things. At this juncture, localities cannot depend on any consistent policy other than one that places more responsibility on communities to develop their own unique solutions for the citizens that reside in them.

The argument advanced in this book is that a nonpolicy, even when there is a shift in emphasis, is a poor and impractical approach to governance. Therefore, it is argued that, *because of the major changes in the international economy—rather than in spite of them—local communities can pursue development policies that complement national economic objectives.* Clearly, larger metropolitan export-oriented economies can take advantage of international development options more readily than can smaller rural communities. Nonetheless, it is important that every community pursue economic policies that, first, enhance or facilitate local industries with international potential and, second, meet the employment needs of all community residents.

The Organization for Economic Cooperation and Development (OECD) gave further strength to the arguments for regional/local economic development within a national policy context (OECD 1986). Its statement proposed that national economic policy must incorporate regional/local economic development to achieve the following goals: to moderate the effect of the pace of the economic adjustment on localities and individuals; to cushion the impact of rapid economic change on firms and affected employees; and, most important as well as appropriate, to revitalize local economies and

Table 2.1   Scheme for Classification of Policies

Sectoral Economic Dimension	Spatial Dimension of Policy		
	None	Implicit	Explicit
No industrial policy	General economic policies	Economic policies with regional implications	Regional policies
Some sectors supported indirectly (e.g., steel, agriculture)	Sectoral policies	Structural policies	Regional development (proper)
Industrial supports			
By sector	Structural policies	Regional structural policies	Local economic development
By firm/plant	Nonexistent	Local business assistance	Firm-/plant-specific structural policies

facilitate adjustment to the economic transformation of the national economy.

The first two goals are to meet short-term objectives of maintaining the standard of social well-being and alleviate the consequences of rapid external economic change on certain areas and groups. Those policies aimed at regional and local revitalization and adjustment include sustained investment, medium- and long-term job creation, and the building of local institutions capable of sustaining the area's economic vitality.

The objectives of regional/local economic development, according to the OECD (1986), include the following:

- Strengthening the competitive position of regions and localities within regions by developing the potential of otherwise underutilized human and natural resource potential
- Realizing opportunities for indigenous economic growth by recognizing the opportunities available for locally produced products and services
- Improving employment levels and long-term career options for local residents

- Increasing the participation of disadvantaged and minority groups in the local economy
- Improving the physical environment as a necessary component of improving the climate for business development and of enhancing the quality of life of residents

Thus, from the national government point of view, the underlying rationale for regional and local economic development policies is to bring about more equitable distribution of development and to take advantage of the enormous capacity of localities to promote and sustain the development process.

The economic performance of localities is inextricably linked to national economic performance. In designing policies to enhance explicitly local economic development, the underlying assumption is that local economic adjustment is a vital component of facilitating sustained national economic recovery. Moreover, it is recognized that local barriers to change, such as the financial or administrative capacity of local government, can be altered so that localities can take part in the economic-development and employment-generation process. In essence, local economic development is both a reaction to larger-scale economic transformation and a positive response to the possibilities of formulating locally based economic solutions in spite of larger-scale economic and political forces.

Local economic development has evolved over the last decade into its present form. No true prototypes exist, because the political-economic framework differs among the nations that practice it. In every sense, local economic development is an emerging field, and the precise form that it will take—when and if it becomes fully developed—will depend on the long-term success of current ventures. At this stage, local economic development is a *movement* rather than a strict economic model from which uniform approaches have emerged.

New Challenges and Opportunities for Localities

If local government and community-based organizations and related institutions take up the challenges related to the national industrial transformations discussed here, they will be increasingly drawn into the debate about which sort of economic action the locality can or should follow. In order to meet this challenge, civic leaders must begin to examine available alternatives in a *realistic* fashion. It may be obvious to local businesspeople that firm declines cause other business failures. It may not be clear, however, whether local leaders have the capacity to formulate plans to deal with local economic problems that cause the decline of the communities' primary industries. Moreover, the issues that communities face differ depending on their circumstances. Nonetheless, almost all American communities will have to come to terms with the following realities.

Few communities can anticipate economic growth based on internal population migration alone. The national population growth rate for rural areas has slowed substantially. Further, the major growth areas are urban fringe communities that lie just outside the large metropolitan cities and remain heavily dependent on those centers for job creation. Population growth in some nonmetropolitan areas is not currently being matched by an increase in employment growth.

It is unlikely that any community will be able to increase significantly their local employment opportunities by attracting new manufacturing firms. The loss in employment among existing firms will probably exceed new manufacturing employment in all regions. As a result, communities and regions will have to become directly involved in cultivating firms from the advanced information and service sectors, mentioned earlier, or in other activities as their new base employers.

Communities based on a single industry (such as agriculture or mining) or on a few major employers will remain more vulnerable than those with a more diverse economic base. As a result, communities with narrow or declining economic bases will have to develop more sophisticated economic strategies to remain economically and socially desirable places.

All communities will feel increased pressure to develop programs that deal with adult long-term unemployment. In addition, new (young) entrants to the labor market, particularly teenagers, will have location-specific employment problems related to job *access* as well as to undereducation and undertraining. Job access issues will arise from inadequate transport, lack of information about jobs through traditional government personnel services and news media, and a lack of knowledge regarding positions available with the new, small, fast-growing firms. Another important factor is the increasing mismatch between job and social attributes of the job seekers.

Comparative geographic or transportation advantage will no longer be determined entirely on the availability of natural resources. Increasingly, the economic development literature indicates that *location* per se, as that term relates to proximity to markets, natural resources, or transportation, is not as important in today's economic circumstances as is the availability of specialized technology-oriented infrastructures. These include research facilities, higher education services, high-quality up-to-date telecommunications, and special financial assistance to accommodate business start-ups or expansions. In essence, the literature suggests that the quality of the support services in the locality determines the potential for new economic activity. Even in a period of industrial restructuring, regions are growing in the West and South. These communities need to plan carefully for both population and economic expansion.

Thus, irrespective of their size or geographic location, localities may be able to construct alternative economic futures by carefully assessing and achieving the best match between their physical, natural, and human resources.

Community Types and Opportunities

Communities of all sizes have an important role to play in helping state and federal government development departments assist new firms in establishing, expanding, and competing for domestic and overseas markets. Those communities that develop the appropriate physical, social, financial, and regulatory environments will be in a better position to take advantage of new forms of industrial and commercial growth. In this context, local economic development opportunities can be undertaken. A community's economic development approach will depend largely on its circumstances. There are three basic types of communities responding to changing economic conditions (see Table 2.2).

First are growing communities with new opportunities for business and commercial growth. These communities are generally on the coastline or in areas adjacent to (within 60 miles of) the large metropolitan cities. Their growth may be part of an urban overspill or people changing location for lifestyle reasons, as a result of, for example, early retirement or dissatisfaction with an urban mode of living. Some of these communities have more resources than others, depending on their population and industrial mix. Clearly, even in these growing areas, not all increases in population translate into new economic activities.

Table 2.2   Development Initiatives by Community Type

Community Type	Development Initiative
Growing area	Manage population growth and plan economic alternatives.
Restructuring area	Diversify existing economic and employment base.
Declining neighborhoods and community	Manage existing public resources. Retain remaining local business. Search for alternative economic activities consistent with government capacity and community employment requirements.

Second are communities facing severe economic restructuring problems. These communities, of which there are many throughout the nation, have lost most or all of their industrial base (e.g., steel or manufacturing). They include both inner metropolitan city areas and nonmetropolitan single-industry communities.

Finally, third are small communities, some in the shadow of regional centers and others in isolated rural areas, with declining economic and sometimes population bases. Economic development may not be a relevant or meaningful activity for this type of community. Nevertheless, even these communities can use their existing resources more creatively to stabilize their local economy and maintain civic services.

In sum, there are aspects of local economic development useful to every community. Local governments and community groups, with their development professionals, must decide on the basis of later chapters of this book as well as other information what role their communities can and should play in determining their economic destiny.

References and Suggested Reading

Blakely, Edward J., and Philip Shapira. 1984. Industrial Restructuring: Public Policies for Investment in Advanced Industrial Society. *Annals of the American Academy of Political and Social Sciences* 475:96.

Bluestone, Barry, and Bennett Harrison. 1982. *The Deindustrialization of America.* New York: Basic Books.

Bradford, C., L. Finney, S. Hallet, and J. Knight. 1981. *Structural Disinvestment: A Problem in Search of a Policy.* Evanston, IL: Northwestern University Center for Urban Affairs.

Carter, Robert. 1984. The Spatial Basis for Economic Development and Adjustment Policies. In *Regions in Transition,* edited by Robert Carter. Canberra, Australia: Department of Local Government and Administrative Services.

Cohen, Steven S., and John Zysman. 1987. *Manufacturing Matters.* New York: Basic Books.

Drucker, Peter. 1991. The Changed World Economy. In *Local Economic Development,* edited by R. Scott Fosler. Washington, DC: International City Management Association.

Goldsmith, William, and Edward J. Blakely. 1992. *Separate Societies: Poverty and Inequality in U.S. Cities.* Philadelphia, PA: Temple University Press.

Goldstein, Harvey A., and Edward M. Bergman. 1986. Institutional Arrangements for State and Local Industrial Policy. *Journal of the American Planning Association* 52 (Summer): 266.

Hanson, B., R. Cohen, and E. Swanson. 1979. *Small Town and Small Towners.* Beverly Hills, CA: Sage.

Harrison, Bennett, and Barry Bluestone. 1988. *The Great U-Turn: Corporate Restructuring and the Polarization of America.* New York: Basic Books.

The Hollow Corporation: A Special Report. 1986. *Business Week,* March 6.

Jacobs, Jane. 1969. *The Economy of Cities.* New York: Random House.

Jarboe, Kennan Patrick. 1985. A Reader's Guide to the Industrial Policy Debate. *California Management Review* 27 (Summer).

Ohlsson, L. 1984. *International and Regional Specialization of Australian Manufacturing: Historical Developments and Implications for National and Regional Adjustment Policies.* Bureau of Economics Contributed Paper No. 1. Canberra, Australia: AGPS.

Organization for Economic Cooperation and Development (OECD). 1986. *The Revitalization of Urban Economies.* Paris: Author.

Reich, Robert. 1991. The Real Economy, *The Atlantic*, February.

Silver, Hilary, and Dudley Burton. 1986. The Politics of State-Level Industrial Policy. *Journal of the American Planning Association* 52 (Summer): 277.

Staudohar, Paul, and Holly Brown. 1986. *Deindustrialization and Plant Closure.* Lexington, MA: Lexington.

3 | The Meaning of Local Economic Development

The Basic Concept

The Clinton administration is deliberately fostering a new and more realistic understanding of economic development. Both the national and local policymakers are being urged to pursue national economic strategies to stimulate their local economies. Moreover, in too many instances, the combination of national economic interest and the motivations of multinational firms does not coincide with the needs or interest of local communities, workers, or disadvantaged groups. In market-driven economies, communities are also marketplaces. As a result, communities must market their resources intelligently and gain *competitive advantages* to create new firms and maintain their existing economic base. That is, communities must use their current human, social, institutional, and physical resources to build a self-sustaining economic system.

Those concerned with this new wave of local economic planning often ask whether this activity represents a new approach or is merely a reformulation of failed "trickle down" policies of the past. The key issue is whether this new version of local or community-based economic development is genuinely better and more effective than past efforts. Can local governments and/or neighborhoods, working together or separately, create new jobs? More important (and perhaps more fundamental), are these new approaches just shifting existing jobs around the nation with giveaways and gimmicks? Is it, in fact, possible to generate more work and more "good" jobs in a technology-based economy? If local efforts can generate employment, are these efforts cost-effective? Or is it inevitable that

local institutions are only playing at the margins of the employment-generation process without making any substantial impact
on the real requirements for employment in a transitional economy? Can the job-formation process be related to the people who
need the work? Is it inevitable that the underclass—the racial
minorities, women, and other disadvantaged persons—will not
share in the benefits of any form of economic development, be it
local or otherwise? Is economic development another code term
for corporate control of community assets? Finally, at what cost to
planning, zoning, and environmental considerations is local employment being pursued?

These are difficult questions. Throughout this book, conceptual,
policy, and programmatic differences respond to these queries and
form a rationale for local economic development. This chapter
provides a link between the existing theories of regional growth
and decline to forge an operational paradigm for local economic
development. The conceptual framework for local economic development emerges from basic development theories. This chapter
will not review all development theories but instead provide an
intersection of the public policies that are the basis for local
economic development.

Defining Local Economic Development

Locally based economic development represents not merely new
rhetoric but a fundamental shift in the actors as well as the
activities associated with economic development. Cochrane, in
discussing local government in the United Kingdom, could easily
be describing the U.S. pattern when he comments:

> It is not difficult to see the period since the late 1970s as one in which
> the links between business and government have begun to be forged
> rather more effectively than the past, as part of the process of moving
> toward an "enterprise" state. The language of welfare has been re-
> placed by the language of growth, regeneration, and public/private
> partnership, particularly in urban areas. (Cochrane 1992, 415)

This enterprise approach is essentially a new process by which
local governments, along with local corporate firms, join forces
and resources to enter into new partnership arrangements with the
private sector or each other, in order to create new jobs and
stimulate economic activity in a well-defined economic zone. The

central feature in locally based economic development is in the emphasis on "endogenous development" policies using the potential of local human, institutional, and physical resources. There are two distinct approaches to forging the partnership.

Robinson (1989) labels them as (1) the *corporate-center approach,* which places the emphasis of economic development on urban real estate development and industrial attraction; and (2) the *alternative approach,* which attempts to steer economic development activities to local disadvantaged residents. These approaches are summarized in Table 3.1.

The focus of economic development in this book is derived from Robinson's alternative view. These strategies, however, must be balanced in any local economic development program. The central feature of locally based economic development is in the emphasis on endogenous development using the potential of local human and physical resources to create new employment opportunities and to stimulate new, locally based economic activity. The reasons for this are summarized as follows:

> [Previous economic development theories and program efforts] . . . have relied too heavily on a belief that the benefits of economic growth and expansion will "trickle down" to improve conditions of the poor. . . . They have separated macroeconomic policies and maintenance programs into two separate and distinct camps; and . . . they have focused almost exclusively on trying to remedy perceived "defects" in the poor—inadequate education or skills, weak community supports, lack of motivation—and ignored the very real, potent barriers in the structure of opportunities the poor confront on the "demand" side of the labor market equation. (Corporation for Enterprise Development 1982, 2)

Local economic development is process-oriented. That is, it is a process involving the formation of new institutions, the development of alternative industries, the improvement of the capacity of existing employers to produce better products, the identification of new markets, the transfer of knowledge, and the nurturing of new firms and enterprises. As Giloth and Meier state:

> Balanced growth as a strategy to ameliorate urban inequities was rooted in the urban landscape of the late 1970s and early 1980s. . . . The pervasive experience of glitter and collapse brought together

Table 3.1   Two Economic Development Policy Approaches and Their Dimensions

Dimension	Corporate-Center Approach	Alternative Approach
Public and private sector	Primacy of private sector market decisions: private sector lead	Private sector market decisions influenced by public sector interventions: public sector
	Public sector responsible for creating an economic and social climate conducive to private investment	Public sector responsible for guiding private investment decisions so they generate desired economic development outcomes
Public sector planning	Objectives favoring growth and tax base expansion	Objectives favoring the creation of direct benefits for low-income and ethnic minority residents
	Planning processes that are relatively inaccessible to low-income and ethnic minority groups	Planning processes that are relatively accessible to low-income and ethnic minority groups
Public sector interventions	Public resources provided as a means of accommodating needs of private industry	Public resources provided conditionally as a means of ensuring specific economic development alternatives
	Intervention in areas likely to generate growth (e.g., attraction of businesses from outside the city)	Intervention in areas likely to produce direct benefits for low-income and ethnic minority residents (e.g., retraining of displaced workers)
	Targeting of growth sectors (e.g., advanced services, high tech, tourism)	Targeting of growth sectors and sectors able to meet important economic needs
	Targeting of headquarters and branch plants	Targeting of locally owned establishments
	Concentration of projects in central business districts and surrounding areas	Decentralization of project locations
	Emphasis on the creation of jobs for white-collar and highly skilled workers	Emphasis on the range of local labor needs, including those of underemployed, unskilled, and blue-collar workers

SOURCE: Robinson 1989, 285. Reprinted by permission of the *Journal of the American Planning Association*.

grass-roots constituencies with a new set of ideas. A long-gestating realignment of neighborhood activism toward economic development and redistributive/accountability strategies was unwittingly spurred by the reduced funding, privatization, and entrepreneurial ideology of Reaganomics. This led neighborhood and civic activists to reformulate notions about development, governance and their role. (Giloth and Meier 1989, 185-186)

Peter Eisenger (1988) provides a conceptual basis for government led entrepreneurship in *The Rise of the Entrepreneurial State*. No matter what form it takes, local economic development as articulated by Giloth and Meier (1989) has one primary goal: to increase the number and variety of job opportunities available to local people. To perform these activities, local governments and/or community groups must take on an initiating rather than a passive role.

In essence, local government, using the resources of existing community-based institutions (where they exist and possess economic potential), must assess the potentials and marshal the necessary resources to design and develop the local economy. Local government and community organizations are realizing that *all* public sector actions have an impact on private decisions. Even the most narrow local governments, those that restrict their activities to the traditional housekeeping services, have affected economic development in their communities, if only through their passivity. Many local governments have probably acted unwittingly to restrict employment opportunities without understanding or assessing the economic consequences of their actions.

Similarly, neighborhood-level community institutions, both non-profit and public, have had dramatic impacts on private investment. Neighborhoods with active churches and neighborhood organizations that work toward the constructive development of their community act as beacons to developers and investors. Further, good schools, both public and private, are an essential factor in the potential location of new businesses. Private decisions and public economic activity are intimately related; both affect employment opportunities for all local residents. This concept should lead local governments and community-based organizations to take a new and different perspective toward planned, coordinated development initiatives. Communities large and small need to understand that, no matter how depressed or wealthy they are, local government, community institutions, and the private sector are essential partners in the economic development process.

Theories of Growth and Development

Currently, no theory or set of theories adequately explains regional or local economic development. There are several partial theories that can help us understand the underlying rationale for local economic development. The sum of these theories may be expressed as:

Local/Regional Development = f(natural resources, labor, capital investment, entrepreneurship, transport, communication, industrial composition, technology, size, export market, international economic situation, local government capacity, national and state government spending, and development supports).

All of these factors may be important. However, the economic development practitioner is never certain which factor has the greatest weight in any given situation.

The Role of Neoclassical Economic Theory

Neoclassical economic theory does not have a significant spatial dimension. Nonetheless, neoclassical models of large-scale economic systems can be applied to the competitive positioning and wealth generation of a subarea of a larger economy.

Neoclassical theory offers two major concepts for regional and local development: *equilibrium* and *mobility*. These concepts state that all economic systems will reach a natural equilibrium if capital can flow without restriction. That is, capital will flow from high-wage/cost to low-wage/cost areas, because the latter offer a higher return on investment. In local development terms, this means that ghettos should draw capital because prices for property and sometimes labor fail to meet the demand of the marketplace. If the model worked perfectly, then all areas would gradually reach a state of equal status in the economic system. Much of this rationale underlies the current wave of deregulation of banking, airlines, utilities, and similar services. In theory, all areas can compete in a deregulated market.

In a similar manner, advocates of neoclassical theory would oppose any community regulations on the movement of firms from one area of the nation to another or even offshore. Neoclassical theorists oppose actions by community groups and local governments that place restrictions on firm locations, such as minority or local equity participation. They suggest that such moves are doomed to fail and disrupt the normal and necessary movement of capital. Moreover, neoclassical advocates argue, as discussed earlier, that there should be no attempts to save dying or uncompetitive firms. Further, they argue that workers who lose their jobs should move to new employment areas as a further stimulus to development in such places.

Among regional and local economic development advocates, there are many detractors of neoclassical theories and the policies derived from them because of the anti-intervention stance. In addition, neoclassical models tell us little about the real reasons some areas are competitive while others fail. Further, the neoclassical framework is generally viewed as antagonistic to the interest of communities as places with a raison d'être beyond their economic utility.

However, there are useful concepts that can be derived from the neoclassical position. First, in a market society, all communities must ensure that they use their resources in a manner that attracts capital. Artificial barriers, inferior governmental bureaucracy, and an absence of a "good business climate" are, in fact, barriers to economic development. Second, communities or disadvantaged neighborhoods can and should argue for the resources necessary to assist them to reach an equilibrium status with surrounding areas. This can be partially accomplished by upgrading commercial properties through local government loans and grants, as well as by offering training and other programs that enhance the value of local labor. These measures can act as inducements to equalize the value of inner-city neighborhoods and other disadvantaged areas with more prosperous places.

Economic Base Theory

As stated previously, communities are socioeconomic systems. As whole systems, they trade with other communities outside their boundaries. Adherents of economic base theory postulate that the determinants for economic growth are directly related to the demand for goods, services, and products from other areas outside the local economic boundaries of the community. In essence, the growth of industries that use local resources, including labor and materials for final export elsewhere, will generate both local wealth and jobs.

The local economic development strategies that emerge from this theory emphasize the priority of aid to businesses that have a national or international market over aid to local service firms. Implementation of this model would include measures that reduce barriers to export-based firms establishing themselves in an area, such as tax relief, transport facilities, and telecommunications. Moreover, firm recruitment and economic assistance efforts would be aimed at supporting or encouraging export-oriented enterprises.

Many of the current entrepreneurial and high-technology strategies aimed at attracting or generating new firms are based on economic base models. The rationale is that nonexport firms or service-providing businesses will develop automatically to supply export firms or the population that works in them. Moreover, it is argued that export industries have higher job multipliers than local service firms. Thus, every job created in an export firm will generate, depending on the sector, several jobs elsewhere in the economy. There are regional economic methods that will test and measure such impacts of firms on the local economy.

The weakness in this model is that it relies on external demand rather than internal need. Overzealous application of base models can lead to a skewed economy almost entirely dependent on external, global, or national market forces. This model is, however, useful in determining the balance between industrial types and sectors that a community needs to develop for economic stability.

Location Theory

According to an old saying of regional economists, there are only three important variables in regional growth: location, location, and location! This statement holds some truth with respect to industrial site development. Firms tend to minimize their cost by selecting locations that maximize their opportunities to reach the marketplace. The old industrial/manufacturing model postulated that the best location was almost always on the cheapest transport link between raw materials and markets.

Other obvious variables that affect the quality or suitability of a location are labor costs, the cost of energy, the availability of suppliers, communications, education and training facilities, local government quality and responsiveness, and sanitation. Different firms require differing mixes of these factors in order to be competitive. Therefore, communities generally attempt to manipulate the cost of several of these factors to become attractive to industrial firms. All of these actions are taken to enhance a *location* beyond its natural attributes.

The limitation of location theory today is that modern technology and telecommunications alter the significance of specific locations for the production and distribution of goods. In many respects, almost any community can now compete as an urban center because transportation cost for the most sophisticated products has been reduced dramatically. Moreover, less tangible variables, such as the quality of community life, now seem to overshadow the obvious advantages of large market or natural resource areas.

The contribution of location theory to local economic development is the realistic parameters it places on the development process. Communities need to ascertain the relative value of their locational attributes in relation to other resources that the area possesses.

Central Place Theory

The basic concept underlying central place theory is called the *hierarchy of places*. Each urban center is supported by a series of smaller places that provide resources (industries and raw materials) and require a central clearinghouse to filter into the world marketplace. Regional development models for rural areas have relied heavily on central place theory to guide resource allocations among country centers, on the thesis that the development of a central country center with larger-scale population would improve the economic well-being of the entire region. The application of central place theory can be observed in projects such as the Tennessee Valley Authority (TVA), Rural Electrification, the Economic Development Administration (EDA), and similar rural service bureaucracies. Each of these organizations attempted to develop a regional economic plan with one or two communities either designated or emerging as regional nodes for development.

Central place theory has relevant applications for both urban and rural local economic development. It is necessary, for example, to differentiate the functions of various neighborhood areas so that they can remain viable centers. Some areas will become region-serving and others will serve only the resident community. Local economic development specialists can assist communities or neighborhoods to develop their functional role in the regional economic system.

Cumulative Causation Theories

Casual observation of the decay of urban neighborhoods demonstrates the basic concept of cumulative causation thesis: The interplay of market forces increases rather than decreases the inequality between areas. As a result, a divergence in regional income is a predictable outcome. Market forces, by their nature, pull capital, skill, and expertise to certain areas. These areas accumulate a large-scale competitive advantage over the rest of the system. Myrdal expounded this theory and described it in the following manner:

> Suppose accidental change occurs in a community, and it is not immediately canceled out in a stream of events; for example, a factory employing a large part of the population burns down . . . and cannot be rebuilt economically, at least not at that locality. The immediate effect is that the firm owning it goes out of business and its workers become unemployed. This will decrease income and demand. In its turn, the decreased demand will lower incomes and cause unemployment in all sorts of other businesses in the community which sold to or served the firm and its employees. . . . If there are no exogenous changes, the community will be less tempting for outside businesses and workers who had contemplated moving in. As the process gathers momentum, businesses established in the community and workers living there will increasingly find reasons for moving out in order to seek better markets somewhere else. This will again decrease income and demand. (Myrdal 1957, 23)

These "backwash effects" prevent low-income neighborhoods from developing the requisite internal capacity for revitalization. However, the growth of prosperous areas tends to feed on itself if the growth-inducing factors remain conducive. As a result, less-well-off areas, be they rural backward regions or inner-city ghettos, tend to send their capital and labor supply to better places without any significant return. For this reason, many advocates of ghetto capitalism propose the movement of jobs into the neighborhoods rather than the movement of people away from their communities in search of work. In addition, the loss of community retail, banks, supermarkets, and commercial establishments continually drains both inner-city ghettos and rural underdeveloped areas of the requisite internal capital for rebuilding themselves.

The community development corporation and rural development centers are one example of creating new institutional arrangements to rebuild underserved communities. These organizations attempt to restore the market and act as capital retainers or capital attractors for areas where market forces are especially weak.

Of course, the weakness of this theory is in its application to small areas, such as urban inner-city ghettos. Obtaining data that shows capital leakage is difficult, even when the leakage is clearly observable. Further, knowing where to intervene in a decaying neighborhood economy is extremely difficult. Do you reestablish banks or supermarkets? Given that ghetto markets are poor in both money and use of funds, how do you accumulate any reinvestment capital? These are very difficult questions that local economic developers need to consider before they embark on attempts to improve very troubled areas. In a sense, every cause is an effect.

Attraction Models

Industrial attraction theory is the economic development model most widely used by all communities. The basic assumption that underlies attraction theory is that a community can alter its market position with industrialists by offering incentives and subsidies. This assumes that new activity will generate taxes and increased economic wealth to replace the initial public and private subsidies. A more cynical view, supported by considerable evidence, is that the cost of such efforts is in fact paid by the workers and taxpayers of the community (Bluestone, Harrison, and Baker 1981).

Communities are products. As such, they must be "packaged" and appropriately displayed. The objective evidence of this packaging can be observed in magazine and newspaper advertisements extolling the virtues of certain places over others. Some regard this mode of economic development with justifiable cynicism. Nonetheless, there is considerable anecdotal evidence that community promotion works and that failure to use it may be a political liability.

A new approach in attraction is the change in emphasis from attracting factories to attracting entrepreneurial populations, particularly certain socioeconomic groups, to a community or area. New middle-class migrants to an area bring both buying power and the capability to attract employers. In addition, recent migrants are more likely to start new firms. As a result, many communities have reassessed their firm attraction efforts and reoriented them toward "people" attraction. This approach has been particularly effective in rural areas where the quality-of-life factor can attract new populations. This, in turn, has led to increased economic growth as a response to both internal demand and new export enterprises created by the new migrants.

Finally, the attraction model underlies some of the current emphasis on "civic entrepreneurism." The notion is that nations, states, and communities can become attractive places for entrepreneurs to flourish. An emerging corollary theory suggests that some localities offer special "knowledge networks" to act as incubators for high-technology firms or inventors. These areas are natural entrepreneurial centers because they develop a certain style or esprit de corps. Route 128 in Massachusetts, the Silicon Valley, and the North Carolina Triangle, as well as some areas of Florida, have gained reputations as innovation centers.

Communities all over the world are beginning to initiate policies and programs to make their area more attractive to investors, firms, new migrants, entrepreneurs, and others. The theoretical basis for this activity is that places can display themselves and offer incentives that give them a *competitive advantage* over other areas with similar resource endowments. The extent to which all these efforts cancel one another out or provide businesses with unnecessary and expensive incentives is a topic of considerable debate.

This approach suggests, however, that no city or neighborhood should hide its virtues "under a bushel basket." Some form of marketing is necessary; the means and the rationale are as important as the desired result in undertaking this mode of development planning because the ends may not always justify the means.

Toward a Synthetic Theory
of Local Economic Development

Existing development theory is an insufficient template for local economic development activities. Therefore, the alternative approach advanced here is a synthesis and reformulation of existing concepts. It serves as a basis for thinking about and taking action within the local economic development context. (See Table 3.2 for a summary of this approach.)

Employment

The major, and sometimes the sole, rationale for communities to engage in active development efforts is to boost local employment. In the neoclassical model, lower wage rates and cheaper costs are sufficient to create employment. This model thus suggests two possibilities for action: to change the quality of the place by providing special locational incentives or to increase the value of the local labor force.

The myriad job training and job development schemes in this country are testimony to the importance of transforming existing labor into a more useful product for existing employers. The enterprise zone is a direct example of attempts to stimulate job creation for a specific population by altering the value of locations.

The goal of local economic development is not to alter but enhance the value of people and places. This conceptual position suggests that employment development is a function of how the community builds economic opportunities to "fit" the human resources and utilize/maximize the existing natural and institutional resource base. In essence, the emphasis shifts from the demand (firm) side of the equation to the supply (labor and natural resources) side in the conceptual framework for formulating development solutions.

Development Base

The economic base model relies heavily on a sectoral approach to economic development. The approach concentrates on transactions within the economic system rather than the failures and inadequacies of the system in which the transactions are taking place. This approach is based on the notion that the local economy must maximize its internal institutional linkages in the public and private sectors.

Local economic development theory starts with a premise that the institutional base must form a major component of both finding the problems in the local economy and altering institutional arrangements. Building new institutional relationships is the new substance of economic development. Communities can take control of their destiny when and if they assemble the resources and information necessary to build their own future. This is not a closed political process but an open one that places local citizens in a position to plan and manage their own economic destiny.

Location Assets

Technology is shattering the traditional view of physical location as the major determinant of development. Firms, even large-scale manufacturing operations, are not as stable as they have been. No one knows precisely what is "footloose" with regard to locational criteria.

Thus, the old view that the availability of transportation and market systems determine a community's economic viability is outdated.

Moreover, whereas heretofore rural communities spent most of their energy in attempting to acquire roads and related infrastructure to promote development, they now find this thrust insufficient. Some rural areas are growing even without such large-scale investment. It does not seem to matter whether a rural community is a designated population growth center or a target area for increased industrial development or for resource exploitation. A rural area's economic opportunities are determined by the quality of the available human resource base and not the land resource.

Location, by itself, is no longer a "pull" factor. In some respects, urban and particularly inner-city locations represent "push" factors. Both firms and people want to avoid these places because their image is unattractive. Crime and associated issues make it difficult to do business in many urban environments. Lack of cultural and educational facilities can retard the development of many rural communities.

The new local economic development model suggests that there are *location-inducing* factors. These factors apply more to the quality of the local physical and social environment than to larger-scale geographic considerations. Moreover, developing a community's recreational, housing, and social institutions is an important determinant of economic viability. Concentrating on building the social and institutional network creates the *inviting environment* for a firm to develop or locate in a community. In essence, if the structure is organized in the correct manner, economic activity will ensue—it will not have to be pursued.

Knowledge Resources

Research resources are the base for economic development in a "knowledge intensive" world economy. In the modern economy, information, more than goods, is exchanged. The development of new information in, for example, biotechnology, computing, and telecommunications, is of enormous value. As a result, the loci of economic innovation and product development have moved from the field to the laboratory.

Major research universities, research institutes, and research units in business and industry are of enormous significance to a local economy. Therefore, localities must develop ways to tap the intellectual resource centers of their region or area. These intellectual

Table 3.2   Toward a Theory of Local Economic Development

Component	Old Concept	New Concept
Employment	More firms = more jobs	Firms that build quality jobs that fit the local population
Development base	Building economic sectors	Building new economic institutions
Location assets	Comparative advantage based on physical assets	Competitive advantage base on quality environment
Knowledge resource	Available workforce	Knowledge as economic generator

resources can be of major assistance in developing new goods and services or in unlocking the potential of existing natural and other resources.

The quality of an area's human resource base is a major inducement to all industries. If the local human resource base is substantial, either new firms will be created by it irrespective of location or else existing firms will migrate there. Therefore, communities must not only build jobs to fit the existing populace, they must also build institutions that expand the capability of this population. Rural communities and inner-city neighborhoods seldom have higher education or research institutions that service them. Indeed, the rural communities and urban neighborhoods seldom consider the need for such resources beyond the teaching function or community problem-solving requirement. Local economic development, however, both now and in the near future, will be dependent on the ability of communities to use the resources of higher education and research-related institutions. Rather than attracting a new factory that may initially employ thousands, a community is better served by attracting and retaining a few small related research labs in leading-edge technologies that will eventually create jobs and stability for the total region.

The Emerging Framework

A new conceptual framework is emerging (Table 3.2) to serve as the parameter for local economic development. It does not enjoy any status yet. The basic tenets of this framework suggest that local economic development is a process that emphasizes the full use of existing human and natural resources to build employment and create wealth within a defined locality.

References and Suggested Reading

Alonso, W. 1972. Location Theory. In *Regional Analysis,* edited by L. Needham. Harmondsworth, UK: Penguin.

Bluestone, Barry, Bennett Harrison, and Lawrence Baker. 1981. *Corporate Flight: The Causes and Consequences of Economic Dislocation.* Washington, DC: Progressive Alliance Books.

Cochrane, Robert, in Harold Wolman and Gerry Stoker. 1992. Understanding Local Economic Development in a Comparative Context. *Economic Development Quarterly* 6 (4): 415.

Corporation for Enterprise Development. 1982, April. *Investing in Poor Communities.* Washington, DC: Author.

Czamanski, S. 1972. *Regional Science Techniques in Practice.* Lexington, MA: D. C. Heath.

Daniels, B., and C. Tilly. 1985. Community Economic Development: Seven Guiding Principles. *Resources* 3(11).

Eisenger, Peter K. 1988. *The Rise of the Entrepreneurial State: State and Local Economic Development Policy in the United States.* Madison, WI: University of Wisconsin Press.

Friedman, J., and C. Weaver. 1979. *Territory and Function: The Evolution of Regional Planning.* Berkeley, CA: University of California Press.

Giloth, Robert, and Robert Meier. 1989. Spatial Change and Social Justice: Alternative Economic Development in Chicago. In *Restructuring and Political Response,* edited by Robert Beauregard. Newbury Park, CA: Sage.

Goldstein, W. A. 1979. *Planning for Community Economic Development: Some Structural Considerations.* Papers prepared for the conference on Planning Theory and Practice, Cornell University, New York.

Hansen, N. M. 1970. How Regional Policy Can Benefit from Economic Theory. *Growth and Change* (January).

Hirschman, A. O. 1958. *The Strategy of Economic Development.* New Haven, CT: Yale University Press.

Hoover, E. M. 1971. *An Introduction to Regional Economics.* New York: Knopf.

Isard, W., and S. Czamanski. 1981. Techniques for Estimating Local and Regional Multiplier Effects of Change in the Level of Government Programs. In *Regional Economics,* edited by G. J. Butler and P. D. Mandeville. St. Lucia, Australia: Queensland University Press.

Myrdal, G. 1957. *Economic Theory and Underdeveloped Regions.* London: Duckworth Press.

Organization for Economic Cooperation and Development (OECD). 1986. *The Revitalization of Urban Economies.* Paris: Author.

Richardson, H. W. 1971. *Urban Economics.* Harmondsworth, UK: Penguin.

Richardson, H. W. 1973. *Regional Growth Theory.* New York: Wiley.

Robinson, Carla Jean. 1989. Municipal Approaches to Economic Development *Journal of the American Planning Association* 55(3, Summer).

Williams, S., Rt. Hon. 1986, August 4. *Local Employment Generation: The Need for Innovation, Information and Suitable Technology.* Paris: Organization for Economic Cooperation and Development.

4 | The Local Economic Development Planning Process

Local economic development is *a process with a product.* This process, a long-term approach to community capacity-building, helps local institutions reorient themselves and improve the economic potential of a given area. It involves more than ascertaining when, where, and how new employment may be stimulated in a given community. In fact, the acquisition of more job opportunities may have little or no impact on the current population. For example, bringing high-tech jobs to a coal-mining community may not assist currently unemployed coal miners though it broadens the local economic base. Therefore, it is important to determine both the short-term and the long-term impact of any proposed development alternative. The 1980s real-estate-oriented economic development aimed at increasing office workers in downtowns illustrates the tragedy of not matching the new jobs to the existing population base.

New firms that are locally or regionally based and/or owned may have a greater impact on the stability and future of a community than branch plants. Similarly, recent research has shown that small innovative businesses may be more labor-intensive than larger plants. In brief, local economic development planning takes hard work, careful analysis, and long-term commitment of resources to achieve well-defined goals. Merely saying that the area needs more jobs is not enough. The community must be prepared to launch specific activities at some risk that will have long-term, positive, planned impacts on the area as a whole.

Phases in Local Economic Development Planning

There are six generally recognized phases of the economic development process. Table 4.1 is a general representation of these phases, which are depicted as sequential. To a large extent, they follow this order, but there is no need to follow these phases slavishly or pass up an opportunity until the ultimate development strategy is in place. However, "ad hoc" economic and employment development efforts may be both wasteful and unfortunate. Several tasks are associated with each of the six main phases within the planning process. A review of these tasks is included in subsequent chapters.

Process Preconditions

Two main preconditions should be observed in the local economic development process. First, the organization or group of institutions responsible for implementing or coordinating the economic change should be involved in determining the process. This may seem obvious, but in too many cases the economic development process is begun by local individuals in service clubs, labor organizations, or other community groups without the active participation of business, labor, and government. As a result, although everyone applauds the initiative, nothing comes of it.

Second, the economic development area or zone of concern should be specified. The economic area must be a unit with internal consistency and cohesion. The area's actual economic configuration should be carefully determined, irrespective of political boundaries. No economy begins and ends with neighborhood or city boundaries. Local economic development, by definition, is a regional enterprise that of necessity involves all communities sharing a common economic linkage. This is usually determined by the laborshed or an equivalent interconnected zone that is physically and economically integrated. The economic development process is the means of planning an approach for the entire economic zone that will maximize the area's total resource base.

The Planning Resource for Local Economic Development

Most of us understand the products of economic growth—more/ better jobs, increased wealth and incomes, increased opportunities

Table 4.1   The Phases and Tasks of the Local Development Planning Process

Phase I	Data gathering and analysis
	• Determining economic base
	• Assessing current employment structure
	• Evaluating employment needs
	• Examining opportunities for and constraints on economic development
	• Examining institutional capacity
Phase II	Selecting a local development strategy
	• Establishing goals and criteria
	• Determining possible courses of action
	• Developing a targeted strategy
Phase III	Selecting local development projects
	• Identifying possible projects
	• Assessing project viability
	−Community −Commercial
	−Location −Implementation
Phase IV	Building action plans
	• Preassessing project outcomes
	• Developing project inputs
	• Establishing financial alternatives
	• Identifying project structures
Phase V	Specifying project details
	• Conducting detailed feasibility studies
	• Preparing business plan
	• Developing, monitoring, and evaluating program
Phase VI	Overall development plan preparation and implementation
	• Preparing project plan implementation schedule
	• Developing an overall development program
	• Targeting and marketing community assets
	• Marketing financial needs

for personal fulfillment, and many of the criteria we normally associate with improving our community. It is important to realize, however, that economic development is also *a process,* whereby a community creates an environment—physical/regulatory/attitudinal— that effects the economic products of more jobs and growth and vitality. By creating the right environment, local governments and many neighborhoods utilize an essential and important set of development resources (Kemp 1992).

The Physical Environment as a Planning Resource

Local governments are normally concerned with the physical environment—physical infrastructure—which is certainly important to business and industry. The private sector usually has both particular and general requirements for a physical environment. Particular needs often include special transportation services or waste-disposal services. In many cases, these forms of physical environment can be "custom made." In other words, local government can provide the special service or facility for a known and defined business or industry requirement, the fulfillment of which is likely to lead to new local jobs.

One of the most important factors influencing locational decisions for new private sector investment is the attractiveness or amenity of a particular area or city, more commonly referred to as *quality of life.*

Many local governments and neighborhoods have been very much concerned about improving the quality of life in their municipalities, but how often have councils regarded a proposed performing arts center as increasing the potential for attracting private economic development? The same question applies to reduced vandalism or traffic congestion. Industry and business regard "livability" as an important locational factor, and local government is in the best position to improve the local quality of life.

Livability can mean different things to different people. Nearly every community possesses natural features, facilities, or simply an aesthetic quality that endears it to certain people. Local governments must identify their quality of life attributes, build on them, and effectively promote them to the business community. Indeed, many local governments have already begun this process.

Many urban and rural communities throughout the nation suffer serious economic problems and desperately need a revitalization of spirit and effort to improve economic circumstances. A revitalized feeling can be created using various mechanisms. For example, small towns can undertake major beautification or improvement projects that create a sense of cohesion and beauty. These projects may be as small as improving street signs and trees or as major as the restoration of the whole civic center. The type of project undertaken should coincide with business improvements and should not merely be physical improvements for their own sake.

The Regulatory Environment as a Planning Resource

Financial incentives and zoning policies are important inputs to the economic development process. Indeed, they are critical to creating an environment for economic development. Many local governments are now undertaking very thorough reviews of their procedures to ensure that "the cost of doing business" in their locales reflects their desire for economic growth. For example, some cities have recently established "one-stop" business or building assistance centers. Such centers are still fairly novel in spite of much publicity regarding their effectiveness.

The Attitudinal Environment as a Planning Resource

The decisions the private sector makes about investment expansion or relocation are not always based exclusively on hard data. In fact, final decisions, particularly between competing options, are most heavily swayed by "gut feelings" or "seat-of-the-pants" reactions. Businesses sometimes choose not to locate in a certain area because residents are known to be hostile or antagonistic to business. The company does not want to place its employees in that kind of community and may instead select the "next best" location, on the basis of the latter's being more receptive and appreciative of the industry. All locations, worldwide, have nearly equal market access, so a community's presentability makes a very big difference to its future.

Often, the developer must defend positions because the local government has adopted a quasi-adversarial role rather than a facilitating one. Developers generally agree on the difficulties of dealing with local neighborhood activists who are uninterested in the economic and social merits of any project. This criticism applies to many areas of a local government administration, including the executive and technical levels. The unanimous and main criticism from developers is delay: delays in presenting applications to council and receiving decisions. The other main criticism applies to conditions—that is, city officials' wanting to "win at all costs" and their reluctance to think beyond petty regulations. Staff should adopt a more positive attitude toward all city planning matters, such as development applications, resubmissions, and rezonings, and should deal expeditiously and reasonably with each one.

As a matter of policy, cities can establish procedures for all development and other related planning applications and make these available as public information. A public chronology of all development decisions would be an effective way to disprove any unfounded or unfair claims by developers of excessive delays. At the same time, it would encourage a more efficient and time-effective approach by city staff. Some communities are assigning the exclusive responsibility for business/industrial development applications to a senior officer in the planning department. This includes the receipt, processing, and approval (or rejection) of applications as well as the monitoring of their progress to their eventual conclusion.

Local government has all the essential resources it needs to affect economic and employment outcomes. Assembling existing resources into a coherent strategy and identifying the necessary external resources, such as specific finances for a particular project or activity, require that councils develop a new sense of their role and responsibility. Several courses of action or options are open to councils as they determine the appropriate way to respond to the unique circumstances of their locality. These are discussed next, as examples of directions that council policymakers must discuss and decide on.

Selecting a Local Economic Development Role

The first planning step any organization interested in local economic development must take is to decide on the role it wants to play in the development process. The stance an organization takes will shape the planning process in which it engages and the tools it can ultimately use to initiate economic development.

Organizations have constraints derived from their current or historic roles or set by their charter. It is always difficult for an organization to go beyond its mandate. Therefore, the role definition must precede any attempt to shape the local economic environment. There are basically four courses of action open to organizations in taking economic and employment development initiatives: to act as the entrepreneur, coordinator, facilitator, and/or stimulator of local development initiatives.

Entrepreneur/Developer

In this role, the organization takes on the full responsibility of operating a business enterprise. Local government or community-based organizations may decide to operate commercial enterprises themselves. Land or buildings in local government control for conservation or for future development can be made available for economic purposes. Local governments may wish to retain commercial land and buildings in public ownership or turn over these resources to local community groups.

It seems apparent that local governments can make greater use of the commercial potential of land or buildings under their control. Beaches, road verges, reserves, and civic centers may be used for a variety of activities that provide jobs. Local governments could take an active approach to identifying and assessing the possible range of commercial opportunities and balancing these against other objectives. Development of a specialization or "theme" for particular localities may increase visitor traffic and therefore the potential for commercial operations.

Community-based organizations are in a good position to run enterprises as employment generators where private enterprise is unable to do so, or to ensure provision of a service where private enterprise is unwilling to take necessary risks. There may also be a case for local governments providing goods and services (and, therefore, employment) for their own operations, particularly where such goods and services would otherwise be imported from outside the area or region. Examples may include hotels, bakeries, and caravan parks in isolated areas, and concrete or crushing plants and nurseries in urban areas.

When a local government has as its objective the redevelopment of depressed industrial or commercial areas in order to increase local trade, employment, revenue, and so on, it can take an active role either individually or in partnership with community-based groups or private enterprise. Use of planning expertise, compulsory acquisition powers, and provision of incentives for relocation or upgrading of premises are some of the options available.

Coordinator

A local government or a community-based group can act as a coordination body to establish policy or propose strategies for an area's development. Because services delivered both by governments and by community and business organizations have a local impact, local councils are increasingly attempting to provide some leadership in the planning and coordination of services within their areas. An extension of this role to economic development might involve community groups in collecting and evaluating economic information (e.g., employment levels, workforce, unemployment, establishments). It might also involve working with other government agencies, business, and community interests to evolve economic objectives, plans, and strategies. Such an approach ensures that all sectors focus their approaches and resources on similar goals and that limited resources are used in more effective ways. This approach can also ensure consistency with state economic programs and strategies so that the local economy receives maximum benefit from them.

Regional tourist development plans, or economic development plans that have been prepared in some areas, represent a possible approach where such plans are developed as joint statements between the three spheres of government and other sectors. Plans developed and imposed by government alone are unlikely to have the same level of commitment or resource input that joint plans will have.

Regional planning bodies with representation from each sector usually work most effectively with governments to produce these plans. A regional approach will normally be more effective because government attention will be focused on regional economies. It will also represent a more manageable level of cooperation between state and local governments. Regional bodies that adequately represent all sectors, and that produce realistic analyses and approaches, are likely to have high credibility with government and thus considerable political influence. Regional organizations formed by local governments are well placed to play an information and catalyst role for their members. Some communities have formed nonprofit organizations that assemble economic data for communities as well as provide a research capacity from which services can be shared.

Facilitator

Some community groups and/or local governments have decided they can best promote development by improving the attitudinal environment in the community or area. This might involve streamlining the development process and improving planning procedures and zoning regulations.

A city or community group may bring together a range of approaches from different functional areas into a policy statement on economic development. This need not involve the commitment of additional resources but rather the provision of a statement of objectives. It would provide a focus for the local government's existing resources and energies and a base for additional programs as and when that course is determined.

Positive use of planning powers may also include establishing employment or development zones and standards that encourage a particular class, scale, or character of development. Although such approaches are often related to environmental conditions, they can also have economic objectives. These opportunity areas then have the potential to be marketed to prospective business clients through direct approaches and advertising in one form or another.

Finally, local council members, as elected community representatives, can advocate local concerns and bring economic problems and opportunities to the attention of higher levels of government. Their advocacy role will be strengthened to the extent that local government can demonstrate community and business sector support for this position and put forward realistic and achievable remedies.

Stimulator

Both community groups and city councils can stimulate business creation or expansion by taking specific action that induces firms to enter or remain in the community. Stimulation may range from developing brochures to actually building industrial estates or small manufacturing workshops.

In some cases, approaches have even included providing industrial buildings. In at least one state, small manufacturing workshops have been built and leased to operators at reduced rents for the first few years of operation. This is an option for local governments in areas where the provision of suitable premises is a problem.

In tourism, a local government may itself promote a particular "theme" or activity in a key venue when private sector action is not forthcoming. Outlets for crafts, craft demonstrations, or a periodic market are some of the possible uses for council-owned premises.

In numerous overseas examples, local governments have provided premises at reduced rents to community-based enterprises and cooperatives to help meet local employment objectives. The course(s) of action a local council decides to take will depend on the local situation. It would be inappropriate for a council not to use its resources in intelligent ways to benefit the total community. Although local development initiatives are not a panacea for local government or the solution to all local problems, they are significant complements to state as well as federal efforts to stimulate economic and employment development. The issue is seldom whether the council should or should not act, but what action to take and how to take it.

Typology of Planning Approaches

There are two conditions that affect the local economic development planning process: the pressures exerted by international and domestic circumstances, and the realization that local economies across the nation are affected differently—that is, some have growing industrial sectors and others are experiencing industrial decline.

What then are the effects of these conditions on community orientation toward economic development? The orientation or expectation of the community shapes its view toward the economic development process. Moreover, localities need to be aware of their orientations toward economic development in order to improve or alter their development strategy. There are two economic development perspectives: responsiveness to external needs or responsiveness to local community needs. For reasons mentioned earlier, responsiveness to external needs characterizes much of prevailing practice, and local responsiveness is identified with a new, emerging (or latent) practice. This sets the stage for proposing a typology of four distinct planning orientations: two prevailing models (recruitment planning and impact planning) and two emerging models (contingency planning and strategic planning) for local economies that are growing or declining (Bergman 1981). The first two models are in response to conditions as they emerge. Recruitment planning represents what is known as a *pre-active* approach to external conditions. That is, the community initiates activities to build or maintain its economic base in response to competitive conditions. Another approach is to wait until circumstances change; that is, *reactive* planning in recognition of the loss of the existing industrial base. The subsequent two approaches are more thoughtful and responsive to the total dimensions of the regional and national economic conditions. Strategic planning is genuinely *proactive* and builds a long-term responsive community system to the conditions a locality faces, whereas the contingency approach is an *interactive* approach that recognizes the need to be flexible to conditions as they emerge. (See Table 4.2 for a summary of planning approaches.)

Recruitment Planning

Recruitment planning is the traditional approach to economic development used by most localities to attract corporate expansion. Public involvement in this style of economic development planning is quite limited. Private sector vitality and initiative tend to displace expressions of local concern for explicit planning or policymaking. Policies associated with this approach tend to operate on the assumption that all business is good for the community. Industrialization of the area is taken for granted as obvious policy. This is generally linked to a tacit understanding that industrialization will be beneficial to the entire community.

Typical planning for this approach includes a wide array of industrial inducements and efforts to enhance the image of the area's "business climate." Because it is the most familiar style of local economic development planning, variants of recruitment planning can be found in nearly every locality in the country.

Impact Planning

Impact planning is a more recent practice that tries to mitigate or reduce the worst effects of industrial losses in a local economy. It derives from a concern for the effects of plant shutdowns in some communities, particularly the impacts on the labor force, and is episodic in nature. This activity, for most of the nation, constitutes the most recognizable public sector planning approach to economic development. The approach relies heavily on the continued availability of federal funds.

The main assumption underlying federal and local impact policies is that action is short-term and only taken in time of crisis. The best example of this type of policy is the nearly $1 billion set aside to rebuild the business base of South Central Los Angeles subsequent to the Rodney King disturbances in April 1992.

This does not allow for flexibility in response and puts too much emphasis on securing external federal funds. Although federal policies are usually explicit, local policies are usually understood implicitly as background assumptions. Nonetheless, few planners would have difficulty in recognizing local industrial, labor, and enterprise policies if they were stated explicitly. In declining areas,

Table 4.2   Planning Approaches

	Responsive Perspectives		Planning Perspectives	
	Pre-active (I)	Reactive (II)	Proactive (IV)	Interactive (III)
Planning				
Model of practice planning model	Recruitment planning	Impact planning	Strategic planning	Contingency planning
Policy				
Industry	Industrialization	Deindustrialization	New indigenous firms	Building on existing firm base
Enterprise types	Corporate adjustment assistance	Government sponsored	High tech/ new tech	Community-based
Development				
Intervention model	Industrial inducements	Government program expenditures	Public-initiated development	Community-initiated development

SOURCE: Adapted from Bergman 1981. Used by permission.

firms close or gradually reduce operations in older, less profitable plants. This strategy is, of course, the mirror image of the industrial growth cycle for growing areas. Local communities can accept the risks associated with corporate modes of decision-making and investment behavior or they can develop their own strategies to deal with corporate decision making. By extension—although it is increasingly contested in most localities—the acceptance of risk involved with corporate market-driven decisions implies that corporations can move such plants and reinvest elsewhere without implied community obligations. Labor force impacts will be borne primarily by relocated workers, their families, and, indirectly, by other workers in businesses whose employment depends on local wage purchasing power. Loss of wages, psychological and physiological distress, higher tax burdens, greater social program expenditures, and continued obligations to amortize industrial infrastructure are included in the costs that all workers in the community ultimately bear.

Development activities are clearly implied by the term *impact planning*. The typical model is a program brought on by sudden, unexpected plant closures and similar employment activity. Development policies and planning in this approach are formed in order to deal with the local impacts of necessary—but equally uncontrolled—acts of corporate mobility. The programs associated with such policies are generally targeted on the real long-term needs of the community.

Two consequences arise from planning solely for impact effects. First, because they rely on national and state programs in response to these needs, planners pay less attention to efforts to plan for the coherent development of their local economies. Directives that specify criteria for triggers, targets, and related aspects of program grantsmanship displace thoughtful planning. Second, the planner's role in redeveloping a local economy is jeopardized when federal funds are withdrawn (as they often are), when localities can no longer afford the costs of standard impact program models, or when the economic development planners have no alternative style of planning to guide them.

Contingency Planning

Contingency planning is an emerging approach that grows out of the ineffectiveness of impact planning in declining areas and from an awareness that planning must anticipate impacts rather than react to them. As plant shutdowns and adverse economic impacts increase, and as civic leaders come to recognize the scale of the problem, some have questioned the assumptions behind the reactive posture of the impact planning model. Because impact planning responds to episodes of economic distress only *after* they happen, actions taken seldom fit into an overall plan for the area. Moreover, efforts to mitigate these impacts are primarily devised to reduce the effects of corporate relocation. Some local officials have now begun to question their past responsiveness to corporate needs.

Repeated episodes of plant closures, bank failures, and real estate debacles have altered the perspective of local officials toward economic development planning. Contingency planning anticipates the worst and best possible outcomes. It can help mobilize an area's resources and inherent capabilities to deflect or accommodate impacts brought on by external forces.

Contingency planning also assesses the strength of all economic sectors and anticipates prospects for plant shutdowns, plans potential economic redevelopment projects, and provides community organizations and leaders with the information necessary to initiate local actions. To carry out these tasks effectively, such planning should be conducted by a local economic development organization or an existing municipal planning and development department. The change of emphasis requires impact program planners to move beyond their "grantsmanship" skills. They must now apply practical knowledge of how their local economy actually operates, analyze key elements of that economy, and design both economic development projects and policies that can be effectively implemented.

Under the contingency planning model, local policies toward industry, labor, and enterprises would respond to local need. Local development policy would attempt to stabilize industrial sectors with the long-term goal of protecting the locality's economic base. Economic development planning is responsible for estimating what a "sustainable" level of industrial activity would be in that local economy. Knowing the structure and linkages among the full complement of remaining industries and their likely tenure, as well as new industrial investment potential and the total need for local resources that can realistically be provided, planners would be expected to help local officials devise industrial stabilization policies geared to the realities of the situation.

Worker buyouts, employee stock option plans, producer cooperatives, worker-community enterprises, community development corporations, and other innovative examples of efforts over the past decade to stabilize job-loss economies are worthy of close examination. Many of these approaches require early public involvement in the planning, financing, and implementation phases. Local labor and enterprise policies should be established *before* the need arises. Here again, economic development requires the study of conditions and circumstances under which others have successfully stabilized local economies and prepared contingency plans for planners' active involvement in each of the phases.

The development activities associated with contingency planning include many of those discussed previously for recruitment and impact planning. But within the contingency planning style of economic development, these activities tend to be community-initiated. It might be said that contingency planning, as a style, began to emerge only after community-initiated efforts demonstrated a clear need for them. Established first as a device for disadvantaged groups to exercise economic power in otherwise strong local economies, community economic development has gone beyond its early concerns to include worker- and community-initiated responses to plant shutdowns.

Strategic Planning

Strategic planning is the most appropriate approach for all communities. This is a future-oriented planning approach that builds a local economy on the basis of local needs. As used here, a dictionary definition of *strategic*—minus its military sense—provides all the essential elements: "utilization of all of [a locality's] forces, through large-scale, long-range planning and development to ensure [success]."

To help ensure the successful development of a stable and prosperous economy, localities adopt a *long-range* view of economic development. This posture avoids the problems occasioned by rapid, almost haphazard growth so prevalent in growing areas. It also allows an area time to organize its capacity to plan for economic development and to accommodate desirable expansion of the economic base properly. Studying lessons learned the hard way in other places is well worth the time, even if local business leaders become a bit impatient with this deliberate approach. A long-term view fosters open discussion and full consideration of plans and policies that affect all segments of the community.

Strategic planning also necessarily entails a large-scale effort to deploy available resources. This does not necessarily mean heavy doses of local government expenditure or large state government subsidies; in fact, it could mean reducing current expenditures from all sources and lowering the risks of heavy long-term public and individual costs. The term *large-scale effort* implies that economic development becomes a long-term objective for all ongoing local community governance functions. The full set of regulations, tax policies, public works, and local government program expenditures is framed with long-term economic development objectives firmly in mind. A strategic view of planning would put economic development specialists at the focal point of budgets, tax policies, public procurement, expenditure patterns, and public finance. Where these functions are now essentially managed with conflicting or no explicit policy direction, strategic considerations would call forth a form of local economic planning. The institutional setting for this strategic guidance would consist of a local economic planning unit or development board placed close to these ongoing municipal functions. The strategic style of planning thus boils down to doing the everyday business of local government with one additional long-term objective firmly in mind: economic development.

Finally, the enterprise types to be developed are selected on the basis of the community needs and resources rather than on the availability of opportunities. In this respect, nearly all actions subject to local decision making would favor no enterprise form in particular. Thus corporations, franchise businesses, small businesses, cooperatives, community-worker enterprises, and all other potential forms of economic enterprise would find themselves on equal footing with respect to local economic development. Such a policy will doubtless require actions by localities to reduce favored treatment of some enterprises and to increase it for others.

The intervention model proposed here is one of publicly initiated development. A very active public role is envisioned, one that orients many public actions toward economic development, attaches a sophisticated strategic planning function to appropriate departments and agencies, and promotes integrated policy positions on the part of major actors. There is evidence that localities are considering the adoption of some of these features, but no place has so far adopted a fully developed version of strategic economic development.

Features of Local Economic Development Planning

Local economic planning is based on a number of key concepts that form the core of the development process. One has to be guarded here because the process is still unfolding and its direction is not yet fixed. Local economic development, as an area of practice, remains a collection of historical activities and reactions to current circumstances. It is not a recognized area of government activity such as planning and zoning or even city administration. Some practitioners of local economic development work for local government in development departments. Others are associated with chambers of commerce or industrial development or attraction offices, and still others are employed by neighborhood- and/or community-based organizations. The variety of circumstances associated with local economic development is still very diverse. Nonetheless, both history and circumstance provide a guide to the salient features of those concepts that are truly developmental.

Zone of Action

This idea is central to economic development and employment planning. It recognizes there is a geography of employment and economic distress. There is a need therefore to direct resources, concern, and energy to specific localities irrespective of what macroeconomic or social policies are pursued. Local economic development programs are designed to intervene in the right place at the right time, affecting both people and place.

Building Community-Level Institutions for Development

A primary ingredient in stimulating local economic activity is the designing of locally relevant institutions and organizations. These groups traverse political, economic, and social barriers in both the public and the private sectors in order to promote development. Economic development institutions range from community promotion and industrial attraction committees to local and community economic development corporations. These institutions share several common features. First, they are inclusive. That is, they attempt to bring into their fold those actors that can effect change. Second, they represent both public and private sector interests. Finally, their mandate is to locate the capital necessary to combine with existing resources as the base for economic development.

Local Ownership Stimulation

Creating new businesses or retaining existing ownership in a community is an important component of local economic development. The reason for this is that local firms form the base for headquarters and use local resources, both human and physical, as the base of their operations. In essence, they are usually good community citizens and contribute to the area's development.

Merging the Resources of the Public Welfare System

Public welfare is both bane and boon for depressed communities. It provides an immediate shock absorber for the disadvantaged. However, years of public assistance and generations of the unemployed are clear indicators that this system has limits to its effectiveness. There are experiments to alter the dependence cycle of welfare by merging welfare and job skills formation projects into the same system. This is being achieved by using welfare payments as wage supports or involving recipients in various economic development programs in an effort to reverse the cycle of powerlessness while employing their labor in community self-help efforts.

Linking Employment and Economic Development Policies and Programs

The goal of local economic development is to diversify the local economy in an effort to provide employment options and opportunities for the existing local human resource base. Clearly, overall economic development must be broader than this, but the presumption built into these efforts is that public resources will be devoted to improving the link between jobs available and people available to work. There will never be a perfect match but there can be better matches.

Building Quality Jobs

Attracting firms is not equivalent to improving the circumstances of the people in the community. Not all jobs represent improvement in a community's capacity. It is important to determine which jobs "fit" the local populace and offer them opportunities to increase their skills to competitive levels, both currently and in the future. Therefore, local economic development, as an intervention in the market system, is aimed at increasing the potential for the local population to have secure jobs and income, which in turn stabilizes the community both economically and socially.

Public/Private Venturing

The hallmark of the U.S. experience in local economic development is the combining of the resources of the public sector and the private sector to attain objectives neither could attain alone. Organizing these resources into the correct formula has been a unique American contribution to the concept of economic development based either on local government or on the neighborhood.

Summary and Conclusion

In sum, the circumstances and character of U.S. political institutions have forged a new and unusual blend of processes and institutions to create a new concept called *local economic development.*

Its key feature is the recognition of the capabilities and resources of local people. It depends on the self-help mentality of the nation.

The national government has been a stimulator, a leveler, a financial resource, and a provider of technical assistance. The national government, for good or evil, has only engaged in industrial and national development policy by default when other approaches have not worked. It has continued to modify market mechanisms rather than create new ones. It has used the existing local institutional base in the form of local government—and more recently in neighborhood-level institutions—by giving them political standing. The national and local governments have steadfastly believed that the cure to any problem in spatial or human allocations of development will result from correcting imperfections in the existing market system. It is these imperfections, both large and small, to which localities must attend in order to build or rebuild their economic base to ensure long-term economic viability.

References and Suggested Reading

Beed, Thomas, and Robert Stimson, eds. 1985. *Survey Interviewing: Theory and Techniques.* Sydney: Allen & Unwin.

Bendavid-Val, Avrom. 1974. *Regional Economic Analysis for Practitioners.* New York: Praeger.

Bergman, Edward. 1981. *Citizen Guide to Economic Development in Job Loss Communities.* Chapel Hill, NC: Center for Urban and Regional Studies.

Butler, Lawrence, and R. E. Howell. 1980. *Coping with Growth: Community Needs Assessment Techniques.* Corvallis, OR: Oregon State University, Western Rural Development Center.

Isard, W. 1975. *Introduction to Regional Science.* New York: Prentice Hall.

Jensen, R., et al. 1979. *Regional Economic Planning.* London: Croom Helm.

Kemp, Roger. 1992. *Strategic Planning in Local Government.* Chicago: American Planning Association Press.

Mahood, T., and A. Ghosh, eds. 1979. *Handbook for Community Economic Development.* Sponsored by the U.S. Department of Commerce, Economic Development Administration, Washington, DC. Los Angeles: Community Research Group of the East Los Angeles Community Union.

5

Analytical Techniques for Local Economic Development

The basic planning approach for local economic development is self-education first, strategy development second, and projects third. Rather than viewing communities only as venues for externally determined economic activity, local development planning sees communities as the essential building blocks of the national economy. The development process begins, therefore, by looking inside the community for the objectives, resources, methods, and personnel available to build the local economic and employment base.

Information and Analytical Requirements for Local Economic Development Planning

Determining both the need and direction of the course of action is the most important aspect of intervention in a local economy. The need to take action is both empirical and intuitive. As Bosscher and Voytek (1990) say in their strategic planning manual:

> Only good analysis enables a community to test its aspirations against the constraints of reality. A good strategy is realistic. It places high priority on those objectives and actions that are feasible and can make a positive difference in the community's well-being. Analysis should be thorough enough to help a community create an action program rather than a wish list.

Changes in the local economy are more subtle than the ordinary citizen can detect. Therefore, sophisticated local methods are required to maintain, strengthen, or assist a community, no matter how large or small, in achieving its economic objectives. Shaffer (1989) provides six basic elements that separate this community economic analysis from other types of analysis. These are (1) the community is the focal point of analysis rather than its subcomponent firms or households or a residual or small fraction of a larger national economy; (2) internally and externally accessible community leadership and initiatives form an integral part of the analytical framework; (3) community resources are important components of the analytical framework; (4) citizen participation is the base for data development, therefore, the type of information and its ease of interpretation are as important as the information itself; (5) community economic analysis is a holistic process that examines the community as an entity but recognizes the different human, physical, and economic dimensions; and (6) citizen attitudes must be explained as both opportunities and constraints in economic development analysis.

The information requirements for local economic development planning can be divided into two segments: (1) socioeconomic base and (2) the community's development capacity. This chapter emphasizes the quantitative techniques of ascertaining a community's economic status and options. It also describes the essential development capacities any community must have to translate its economic options into development programs. The Chapter 5 Appendix illustrates how all these factors are used to construct a community economic development strategy.

Socioeconomic Base Analysis

To start the economic development process, information is needed in five broad subject areas.

Area Demographics. The hierarchy of age groupings should be obtained, with particular emphasis on the employment, age, income, and occupational distribution of the population over the last decade; analysis of dependent population should be included by examining the geographical distribution of persons aged under 15 and over 65 years. The purpose of population analysis is to determine the economic vitality of the community and to assess the target population for proposed economic activity.

Labor Market Conditions. This should include information on employment distribution by sex in each industrial category, using standard industrial codes (SICs). Information on the rate of unemployment and underemployment by industrial sector over the last 5 years, at least, should also be included. Employment patterns in the community will show what human resources are available or required for economic development and what portion of the existing labor force requires assistance.

Economic Characteristics. An area's economic base, its changes, and its responses to new economic conditions should be explored; in addition to the past and current situation, factors influencing economic vitality also need to be assessed. When analyzing the community's overall economic characteristics, one must remember that the current structure of the local economy is neither necessarily the most preferable nor is it predestined to continue in the future. Understanding the local economic structure is the essential first step toward designing a long-term economic development program that builds on the opportunities in the community and addresses local employment needs.

Physical/Locational Conditions. The physical features of the area that relate to the economic base should be examined. These include assessments of physical resources (agriculture, mining, timber, and so on), industrial land availability (both serviced and unserviced), transport and communication links, housing stock levels, and also those assets that could be used as tourism venues. Documenting locational assets (and liabilities) assists in identifying the locality's economic advantages (and disadvantages).

Community Services. The range of social, educational, recreational, and cultural services available to a community should be discovered. These add to the attractiveness of the locale as a place to live and work.

The set of categories above is not meant to be inclusive but to provide a base or reference point for building relevant information on the community. Other configurations of data are also useful depending on resources and the circumstances. The set of data categories selected is to some extent a matter of imagination and preference and can always be revised. One rule must apply, however: Taken together, the categories must reflect the relevant whole—that is, a picture of the community economy. At the conclusion of a community socioeconomic base study, it will be possible to summarize a community's economic and employment challenges. The points that might be covered in the summary are as follows:

Population

- The major characteristics of the local population
- Particular assets of the population
- Weaknesses in the population profile
- Potential areas for improving population profile, such as in-migration of certain age or skill groups
- Other assets of or constraints posed by the population

Employment

- Strengths of the existing labor force
- Weaknesses of the current labor force
- Employment requirements of the current labor force and future trends (speculation)
- Education and training requirements
- Special groups requiring employment attention

Economic Structure

- Current pattern and trends of the local economy
- Weak sectors in the economy
- Economic diversification opportunities and requirements
- Economic growth areas

Community Factors

- Strongest community assets
- Assets requiring attention
- Cohesiveness of community leadership

Physical/Locational Conditions

- Major physical resource assets
- Strengths of geographic situation and locational conditions
- Weak aspects of local physical conditions

Types of Data Collection and Analysis

There are three general types of data collection and analysis efforts. The first of these involves tapping existing data records held by local planning or service agencies as well as local personal knowledge. Within the community, there are persons who have experienced the place in every conceivable way and from every conceivable angle—often over great periods of time. People who

live, work, and play in the community can be involved in contributing their knowledge and ideas to the local economic development
planning process through personal interviews, public hearings, and
neighborhood meetings; church, fraternal, and trade organizations;
the press, radio, and television; and other channels.

Beyond this, special efforts should be made to invite community
leaders, as well as other interested and knowledgeable citizens, to
participate in advisory committees and other formalized groups
associated with the planning process. These people often reflect
the collective thinking and knowledge of citizens who share perspectives or expertise but who cannot actively participate. Another type of data collection and analysis effort involves special
studies. The evaluation of performance conducted toward the end
of each planning cycle may be viewed as one such form of special
study. Existing reports, case studies, feasibility studies, and the like
must also be considered.

Second, surveys are another useful method to fill information
needs or to obtain information on a specific topic. Such surveys
may be undertaken by consultants, the planning staff, advisory
committees, local organizations or local individuals with specialized expertise, university students as part of their studies, or
anyone else with the appropriate credentials and credibility in the
community. It is important that whoever is assigned to conduct a
special survey should be constantly aware of the "client." This
means that any printed materials produced should be readily intel-
ligible to a wide audience and must address the needs of the
planning process rather than the interests of the investigator.
Before any study is undertaken, there should be clear agreement as
to the precise scope and duration of the work. This should include
benchmark task-completion dates and frequent progress-review meetings with those responsible for overseeing the project (i.e., the
planning staff and advisory committees).

A final type of data collection and analysis effort involves the use
of aggregate quantitative analysis. This approach is especially useful for identifying relative strengths and weaknesses of the local
economy and for focusing on major problems and opportunities
that become visible only against the backdrop of aggregate economic trends. Such analysis provides a relative status assessment
that can be most useful in formulating goals and objectives and in
identifying and assessing alternative strategies and projects. Unfortunately, the results of these analytical techniques are often presented
in a manner that makes them mysterious and even threatening. There

are, however, rather simple ways to use them effectively. Several of these methods are described in detail in this chapter. This chapter is not intended to provide a definitive view of quantitative techniques. Several other works provide in-depth discussions of analytical approaches. Among the most relevant works for local economic development professionals are Shaffer (1989), Bosscher and Voytek (1990), and Bendavid-Val (1980).

The Goal of Data Collection and Analysis

Although the details of data gathering and analysis for local economic development planning will vary between local government areas, the basic tasks to be performed will in most cases be the same:

- To determine the climate for economic development in the jurisdiction, collect data on economic and political conditions, and decide on goals
- To determine which agencies currently working in economic development are accessible to local government for the purpose of coordination and planning
- To determine whether economic development actually should be conducted by local government alone, or in coordination with another agency, or by another agency alone
- To identify the barriers to coordination that exist in the jurisdiction, if coordination is desired
- To develop support for the proposed economic development activities from community groups, political groups, and labor unions, among others (e.g., by establishing a formal advisory council with representatives from agencies involved in coordination)
- To develop a set of milestones, a means of monitoring progress toward achievement of the milestones, and a plan for revision of activities according to observed results

The methods selected should be based on the goal set, rather than determining the goal after the data are available.

Regional and Local Quantitative Methods

In order to alter the economic capacity and direction of the locality, one has to know the current condition and the way any change might influence the local economy. There are a variety of spatial economic methods that provide this type of information. All of these tools are controversial because they provide only indirect and oblique measures of the local economy. Moreover, most of these measures were devised for national or large regions rather than for small areas. Therefore, one must be careful about what interpretations are derived from the methods used to assess a local economy's potential and its current or predicted output. Nonetheless, until other measures are developed or tailored to meet small-scale economic systems, these remain useful techniques in spite of their acknowledged imperfections.

In this chapter, the community is viewed as an integrated economic unit that performs like a national economy. That is, the local economy is depicted as producing, importing, and exporting goods and services both for itself and for other areas. Of course, the logic of this model is more complete when applied to a state or large substate region than when used to assess the economic capacity of small subareas of a major urban metropolitan economy or of neighborhoods. These analytical techniques serve as a proxy for local economies. A local community is viewed as a complete economic system, as depicted in Figure 5.1.

Healthy economies, as shown in the figure, export goods, retain substantial income, and build internal linkages that create jobs for local individuals. That is, in examining a community, we are looking for the relationships among those components of the local economy viewed as basic or exporting, that provide local services and increase (multiply) the job and income possibilities for local residents. Determining the size and performance of the local economy with respect to these matters, as well as the community's ability to alter its relationship to other areas in the region, is the focus of these analytical techniques (Landis 1986; Rochin 1986).

Measures of Economic Growth and Linkage

Measures of economic linkage basically describe the relationship between the local economy and its neighboring environment. The most useful techniques provide comparative measures or status descriptions of the local economy.

Shift Share Analysis

Shift share analysis is a powerful and useful technique for analyzing changes in the structure of the local economy in reference

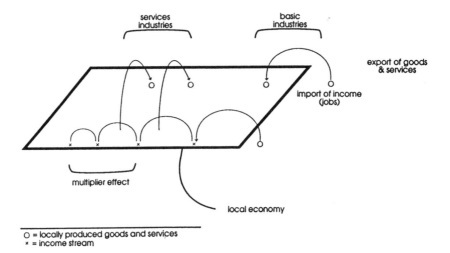

Figure 5.1. Model of Local Economy

SOURCE: Adapted from Agajanian, 1987. Used by permission.

to the state or nation. The community under study can be small or large as long as it is contained in the reference economy. The reference economy can be as small as a county or as large as the nation. As the reader might anticipate, this type of analytical technique is not useful at the neighborhood level.

The purpose of shift share analysis is to determine the job performance or productivity of the local economy in comparison to a larger base (region, state, or nation). Shift share analysis provides data on the performance of the local economy in three related areas:

1. The *economic growth* measures the economic growth in terms of the total employment within the reference economy between the two periods.

2. The *proportional shift* measures the rate of growth of the individual sectors as compared with the rate of growth of the total reference economy. This measure allows one to ascertain whether there are changes in the reference economy.

3. The *differential shift* assists in determining how competitive local industries are in comparison to the reference economy. It measures the rate of growth of the local industrial sectors as compared to that of the same industry in the reference economy. Thus, a positive

differential shift in a particular industry indicates that it is growing faster than the same industry in the reference economy.

Data Requirements for Shift Share Analysis

Shift share analysis requires employment data from two reference years for both the local economy and the reference economy. These data are available through a variety of sources. The most widely available sources for this information are *County Business Patterns*. Information must be developed by standard industrial codes (SICs) for each sector or subsector under analysis.

Shift share analysis can be expressed as:

$$\text{Employment changes in local industry i 1980-1990} =$$
$$\{\text{ref90/ref80} - 1.0\} + \{\text{emp90i/emp80i} - \text{ref90/ref80}\}$$
$$\text{(economic growth)} \qquad \text{(proportional shift)}$$
$$+ \{\text{loc90i/loc80i} - \text{emp90i/emp80i}\}$$
$$\text{(differential shift)}$$

where ref80 = 1980 employment in reference economy; ref90 = 1990 employment in reference economy; emp80i = 1980 employment in industry i in the reference economy; emp90i = 1990 employment in industry i in the reference economy; loc80i = 1980 employment in industry i in the local economy; and loc90i = 1990 employment in industry i in the local economy.

Calculating Shift Share

Shift share analysis is relatively simple to calculate, particularly when using microcomputer technology. Several software templates developed for this purpose are available. In this example, the community of Riverside has 10 sectors.

Illustration of Shift Share

Assuming the economy we are examining has 10 sectors, a shift share analysis might look like this:

	Riverside County Employment (000s)		Economy Growth	State Employment	
	A	B	C	D	E
Sector	1980	1990	(b-a)/a	1980	1990
Agriculture	3,694	6,184	0.6741	55,441	99,427
Mining	1,523	727	−0.5227	46,587	41,070
Construction	13,878	39,719	1.8620	495,645	727,717
Manufacturing	28,647	40,664	0.4195	2,100,467	2,167,133
Transportation and utilities	9,987	12,672	0.2688	528,861	625,508
Wholesale	8,259	13,478	0.6319	566,268	812,777
Retail	41,142	75,446	0.8338	1,679,672	2,276,880
Service	36,003	83,595	1.3219	2,058,321	3,608,896
Finance, insurance	10,178	18,111	0.7794	666,218	893,430
Nonclassified	2,296	1,435	−0.3750	77,830	38,678
Total	155,607	292,031	0.8767	8,275,310	11,315,516

1. The economic growth factor is calculated by subtracting 1 from the ratio of the 1990 total employment and the 1980 total employment.

2. The proportional shift is calculated for each sector by subtracting the ratio of total employment growth in the reference economy from the ratio of sectoral employment growth in the reference economy. Positive entries mean that the sector grew more rapidly than the reference economy. Of course, negative entries are the opposite.

3. The differential shift is calculated by subtracting the ratio of sectoral employment growth in the reference economy from the ratio of employment growth in the local economy. Positive entries indicate a strong competitive position.

Sector	F Economic Growth	G Proportional Shift	H Differential Shift	I Total F + G + H
Agriculture	0.3677	0.4256	−0.1193	0.6741
Mining	0.3677	−0.4862	−0.4042	−0.5227
Construction	0.3677	0.1005	1.3938	1.8620
Manufacturing	0.3677	−0.3360	0.3877	0.4195
Transportation and utilities	0.3677	−0.1339	0.0351	0.2688
Wholesale	0.3677	0.0676	0.1966	0.6319
Retail	0.3677	−0.0122	0.4782	0.8338
Service	0.3677	−0.8708	0.1280	−0.3219
Finance, insurance	0.3677	0.3856	0.5686	0.7794
Nonclassified	0.3677	−0.8708	0.1280	−0.3750
Total	0.3677	0.0000	0.5090	0.8767

Location Quotients

The location quotient is a technique used to augment shift share. It helps to ascertain the export capacity of the local economy and the degree of self-sufficiency of a particular sector.

Data Requirements for Location Quotients

The same basic data sources provide the data for location quotients as for shift share analysis. The basic data can come from any sources that provide information on employment by SIC code for a recent period for both the local economy and the reference economy.

Calculating the Location Quotient

The location quotient assumes that there is little variation in spending patterns geographically, that labor productivity is uniform, and that each industry produces homogeneous goods in the sector. The formula for these calculations is as follows:

$$\text{Location quotient} = \frac{[\text{sect. empl/tot emp}]}{[\text{Ref. Sect. Empl./ Ref. Tot Emp}]}$$

where sect.empl = local sector total employment; tot emp = total employment in all local sectors; Ref.Sect. Empl = reference economy sectoral employment; Ref.Tot Emp = total employment in the reference economy. Using the data from above, the location quotients for sectors in Riverside County are as follows:

Sector	Riverside Employment 1980	State Employment 1990	Location Quotient
Agriculture	6,184	99,427	2.41
Mining	727	41,070	0.69
Construction	39,719	727,717	2.12
Manufacturing	40,664	2,167,133	0.73
Transportation and utilities	12,627	652,508	0.75
Wholesale	13,478	812,777	0.64
Retail	75,446	2,276,880	1.28
Finance, insurance	18,111	893,430	0.79
Services	83,595	3,608,896	0.90
Nonclassified	1,435	38,678	1.44
Total	292,031	11,318,516	1.00

As can be seen, Riverside is relatively healthy in most sectors of its local economy. Some additional analysis of sector A might be undertaken, depending on a general impression of the function of that sector in the local economy.

Scalogram/Hierarchy Analysis for Neighborhoods and Small Towns

Measuring the soundness and determining the structure of small economies such as rural towns, or sub-economies such as neighborhoods, has presented local economic development specialists with considerable difficulties. The regional techniques described above are inappropriate or unusable. Therefore, other methods have to be used that inform us about the comparative status of these economies without the necessity of resorting to expensive local surveys. In the 1960s, a set of descriptive methods was devised to fill this vacuum. These methods are based on various forms of scaling techniques based on settlement hierarchy theories.

Without going into all of the theoretical basis for hierarchy analysis, the ideas can be explained very easily. The basic notion is that human settlements tend to be organized into a hierarchy ranging from the smallest basic service centers tied to higher-level market towns, which in turn are linked with urban centers. Each of these communities has sets of services available related to its role in the hierarchy. Thus the smallest, most isolated centers have a few basic functional activities catering to the immediate needs of the local population, such as grocery stores and gas stations, with the higher-order centers providing a more diverse set of retail as well as social and recreational services (Young and Young 1973).

A derivation of hierarchy theory holds that, when size is controlled, there are still noticeable and documentable differences in the social and economic structures of different places. Structural analysis using hierarchy scales called *Guttman scales* provides a useful technique for obtaining empirical data at the neighborhood or small community level. The analytical device employed is a simple notation of the presence or absence of certain types of socioeconomic activity. This is not the total number, but an indicator of the relationship of that activity to others within the same geographic area. Thus, for example, if the analyst observes a supermarket in a community, there is almost certainly a gas station and restaurant in the same vicinity. However, a community that has a gas station might not have either a supermarket or a restaurant.

Structural analysts have subdivided the areas of analysis into three classes, with the unit of analysis always a single representative institution of some generic type. One class of analysis is based on the economic system (differentiation) of the area observed, another on the social institutional base (solidarity), and the final class is the political-economic base (centrality). Each of these classes is described more fully below.

1. *Differentiation* is defined as the complexity of the neighborhood's economic structure. That is, the number and type of commercial establishments are used as a proxy for the degree of economic strength of the area and its ability to remain an attractive place to live and work.

2. *Solidarity* is an attempt to measure the social activism of an area indirectly. The number and range (social diversity) of the type of voluntary social service and nonprofit neighborhood or community-based institutions illustrate community cohesion. This measure is the same as that described for an economic institution; however, the unit of analysis is the social service institution rather than a firm. The notion is that the broader the social service spectrum, the richer the problem-solving capacity of the community.

3. *Centrality* is a measure of how well connected the community is. Its municipal system and the external political-economic networks give the area increased identity within a group of rural communities or distinction among urban neighborhoods. Centrality is measured by developing a hierarchy of civic institutions such as post offices, high schools, city offices, radio stations, and similar citywide or areawide functions located in the neighborhood. Again, a diverse set of civic activities is an indicator of the significance or centrality of the community to the city or region (MacCannell 1979, 46).

Figures 5.2, 5.3, and 5.4 are meant to illustrate this analytical technique for urban neighborhoods. The value of hierarchy analysis is in allowing laypersons to become involved in neighborhood comparative research without expensive data-gathering methods. In addition, it provides an intuitively correct model for neighborhood-level planners to ascertain where they need to intervene in correcting imbalances in the local fabric of the community before they launch large-scale development schemes that might overwhelm nonprofessionals. Although the measures are imperfect, they are nonetheless instructive.

	Nob Hill	Fruitvale	Adams Point	Mosswood	Dimond	Russian Hill	Southside	Northeast I	North Beach
City organization	1	1	1	1	1	1	1	1	0
Local office of national organization	1	1	1	1	1	0	1	0	0
County or regional organization	1	1	1	1	0	0	0	0	0
Local office of state organization	1	1	0	0	1	0	0	0	0
Headquarters of state organization	1	0	0	0	0	1	0	0	0
Headquarters of national organization	1	0	0	0	0	1	0	0	0

Figure 5.2. Neighborhood Scale of Centrality

As stated earlier, these measures are indirect—some would even say crude. A final caveat: The predictive value of these techniques to support interventions needs to be better documented before they can be used as the sole tool for analysis at the local neighborhood level.

Output Measures

Economic activity does not always equal economic development. Therefore, other measures have been devised to determine the potential effects of certain types of economic activity on the community. These measures are both qualitative and quantitative. Some economic activities, as I have discussed under viability analysis, may not fit the needs and/or interest of some communities. These problems may be uncovered in a social or environmental impact statement. Other quantitative measures provide some guides to the most likely income and job consequences of a development for a city. Of course, these measures are imprecise because workers commute from very wide-ranging areas to their places of employment. Nonetheless, even with this qualifier, this information is useful.

	Fruitvale	North Beach	Adams Point	Southside	Paradise Park	Dimond	Nob Hill	Russian Hill	Castlemont	Northeast I	Mosswood	Belding Woods
Youth recreation center	1	1	0	1	1	1	1	0	0	0	0	0
Senior services	1	1	0	1	1	1	1	0	0	0	0	0
Children services (including day care)	1	1	0	1	1	0	0	0	0	0	1	0
Advocacy services	1	0	1	1	1	0	0	1	0	0	0	0
Youth services/ counseling	1	0	1	1	0	0	0	0	0	0	1	0
Adult education	1	1	1	0	0	0	0	1	0	0	0	0
Employment and training	1	1	1	0	0	0	0	0	0	1	1	0
Legal services	1	1	0	0	0	0	0	0	0	0	0	0
Mental health	1	1	0	0	1	0	0	0	0	0	0	0
Community health clinic	1	1	0	0	0	0	0	0	1	0	0	0
Community organizing services	1	0	0	0	0	0	0	0	0	0	0	0
Library	1	0	0	0	0	1	0	0	0	0	0	0
Food bank	1	0	0	0	1	0	0	0	0	0	0	0

Figure 5.3. Neighborhood Scale of Solidarity

Income Multipliers

The income multiplier is the next best approximation of the wealth-generating potential of new economic activity introduced into the community. The basic assumption is that a change in the production sector will result in a local improvement in incomes throughout the community. A simplified version of estimating income multipliers is presented here; more sophisticated methods are available and cited in the references.

	NOB HILL	NORTH BEACH	FRUITVALE	RUSSIAN HILL	DIMOND	NORTHEAST I	ADAMS POINT	BELDING WOODS	SOUTHSIDE	MOSSWOOD	CASTLEMONT
Local Grocery	1	1	1	1	1	1	1	1	1	1	1
Gas Station	1	1	1	1	1	1	1	1	1	1	1
Cleaners & Laundry	1	1	1	1	1	1	1	1	1	1	1
Bar	1	1	1	1	1	1	1	1	1	0	1
Liquor Store	1	1	1	1	1	0	1	1	1	1	1
Restaurant	1	1	1	1	1	1	1	0	1	1	1
Auto Repair	1	1	1	1	1	1	0	1	1	1	1
Drugstore	1	1	1	1	1	1	1	1	1	1	0
Used Merchandise	1	1	1	1	0	1	1	1	1	1	0
Mens & Womens Apparel	1	1	1	1	1	1	1	0	1	1	0
Barber Shop	1	1	1	1	1	1	1	1	1	0	0
Supermarket	1	1	1	1	1	1	0	1	1	0	0
Florist	1	1	1	1	1	1	1	1	0	0	0
Furniture Store	1	1	1	1	0	1	1	1	0	0	1
Hair Stylist	1	1	1	1	1	1	1	1	0	0	1
Travel Agency	1	1	1	1	1	1	1	0	0	0	0
Delicatessen	1	1	1	1	1	1	1	0	0	1	0
Dentist	1	0	1	1	1	1	1	0	0	0	0
T.V. & Electronics Sales & Repair	1	1	1	1	1	1	1	0	0	1	0
Ice Cream Parlor	1	1	0	1	1	1	0	1	0	0	0
Locksmith	1	1	0	1	1	1	0	0	0	1	0
Jewelry Store	1	1	1	1	1	0	0	0	0	0	0
Tailor	1	1	0	1	1	0	0	0	0	0	0
Health Food Store	1	1	0	1	1	0	0	0	0	0	1
Attorney	1	1	1	1	1	0	1	0	0	0	1
Printing Shop	1	1	0	1	1	0	0	0	1	1	0
Hotel	1	1	1	1	0	0	0	0	0	0	0
Meat market	1	1	1	1	0	0	0	1	0	0	0
Podiatrist	1	0	1	1	0	1	0	0	0	0	0
Shoe Repair	1	1	1	1	0	1	0	1	0	0	0
Game House	0	1	1	1	0	0	0	1	1	0	0
Theater	1	1	1	0	0	0	0	0	0	0	0
Tax Service	1	1	1	0	0	0	0	0	0	0	0

Economic
Diversity
High ⟵——————————————— Low

Figure 5.4. Neighborhood Scale of Economic Differentiation (Diversity)

Data Requirements for Calculating Income Multipliers

Income multipliers are best calculated by doing an economic base analysis that reveals the current amount of money spent locally on various goods and services. In some instances, specific studies have been made for this purpose. It is possible to use standard information on spending ratios related to incomes as developed by the U.S. Census Bureau. In addition, there are a variety of estimating techniques that have been developed by marketing firms.

Calculating Income Multipliers

The formula for calculating the income multiplier is:

$$k = \frac{1}{1 - (\text{MPC1} \times \text{PSY})}$$

MPC1 represents the proportion of local income spent within the locality. PSY is the proportion of local spending translated into local income. In other words, some of the spending leaks out of the locality because corporate headquarters are located elsewhere or because of local spending patterns. Calculating the spending and retention of income for Riverside takes some educated or informed guessing. For example, if we estimate that Riverside will add a new plant that can increase the local economic activity via its payroll by a substantial amount, we might estimate MPC1 as .35 on the basis that 35% of local income is spent locally, and similarly estimate PSY as .45 on the basis that 45% of the local spending is retained in the form of income, then our formula would look like this:

$$\frac{1}{k - (\text{MPC1} \times \text{PSY})} = \frac{1}{1 - (.35 \times .45)} = \frac{1}{1 - .15} = \frac{1}{.85} = 1.18$$

Interpretation

This means that the hypothetical increase in economic activity predicted for Riverside in this analysis is only about 18 cents for every new payroll dollar.

Employment Multipliers

The purpose of this multiplier is to develop the number of new jobs created by new economic activity in the community. This multiplier, like the one above, depends on other estimates or indicators.

Data Needed

The basic information needed for calculating the employment multiplier is the following:

1. Location quotients for all segments of the local economy (SIC) codes
2 The total employment in each sector of the local economy by sector

Calculating Employment Multipliers

The formula for calculating the employment multiplier is:

$$\text{Employment multiplier} = \frac{\text{Total employment}}{\text{Export employment}}$$

In order to calculate this, it is assumed that Riverside has added a new plant that provides employment in the manufacturing sector. For the purpose of this exercise, the Riverside economy is divided into manufacturing and trade.

Sector	Location Quotient 1990	Employment 1990
Manufacturing	4	1,500
Trade	2	1,832

The next step is to determine the portion of each sector that is serving the export market. This formula is $(1 - 1/LQ) \times 100$. Therefore, for manufacturing, the portion of employment serving the export market is $(1 - \frac{1}{4})100 = 75\%$. This means that 1,125 workers out of the total 1,500 serve the export market. For trade, the same formula yields $(1 - \frac{1}{2})100 = 50\%$, or 916 employees in the export sector. If we add these, we obtain 2,014 employees in the export arena. Then, using the formula above:

Employment multiplier = Total employment / Export employment

$$\text{or } 1.63 = \frac{3,332}{2,041}$$

Interpretation

The employment multiplier of 1.63 suggests that for every 100 manufacturing jobs we will see 163 total jobs with 63 in sectors not associated with the manufacturing plant.

Input-Output Analysis

Specialist skills are required to use this form of analysis. It should only be employed for situations in which such information will provide better guidance for the decision-making process and not merely to use fancy techniques not well understood by lay people. A case illustration of how these methods are employed to develop an economic strategy is included as a case study at the end of this chapter.

Input-output (I-O) is a regional economic measures technique. It is seldom of value at the local level, so it will not be described in depth here. Usually, I-O analysis involves extensive computer modeling. It allows more sophisticated analytical techniques for developing information on the impacts of increased or changed economic activity on the various constituent parts of a local economy rather than in the aggregate, as in the case of income and employment multipliers.

The degree of disaggregation of the regional economy is dependent on both the data and the amount of money the region has available to collect and analyze information. The most complete I-O data are generated by collecting primary data from firms individually. *County Business Patterns* and the Economic Census of the Department of Commerce can be used to provide secondary data for I-O analysis.

Input-output identifies the interaction or flow of dollars between various segments of the regional economy. A matrix is constructed from the sectoral data that shows the flows between sectors.

Input-Output Multipliers

Simple mathematical transformations can be made on the flows matrix to derive multipliers for each sector. This is the predictive component of the multiplier. Using these measures, such factors can assess output of employment—household income—based on various assumptions. The literature at the end of this chapter provides a more complete guide to this form of analysis.

Economic Capture

One of the most important aspects of local economic analysis is determining what can be done to affect the local economic situation.

Communities are interested in capturing their share of economic wealth either from the internal market or from the wealth of the region. There are several valuable analytical techniques for assessing commercial sectors of a local economy that are useful at the neighborhood or local jurisdiction level. These techniques are frequently and quite correctly identified as *gravity models*. That is, they are based on the notion that funds will flow to the commercial market based on its locational advantages.

Population-Employment Ratio

One of the best ways to determine how well each sector of the local economy is capturing employment opportunities is to determine the proportion of jobs produced for the local population by the economic sector. This type of analysis is called the population-employment ratio. It is relatively easy to calculate.

Data Needed

This is a comparative analytical technique so it is necessary to obtain the same information on a set of communities in the same region or to have similar aspects such as size or locational features in common:

1. Population for each community in the comparative group (this data can be obtained from the state department of finance or revenue)
2. Number of employees in each sector of the economy (SIC) being examined (this information can be obtained through Dunn and Bradstreet or the state department of employment or job services)

Calculating the Population-Employment Ratio

The population of the community is divided by the number of people employed in a particular sector.

Population employment ratio =
Population of city/Number of employees in sector

For example:

City	Population	Employment in Food and Beverages
Riverside	7,052	234
Forestville	6,758	123

Booneville	7,826	175
Mountainpoint	3,246	71
Porterville	7,833	108

Based on these data:

$$\text{Riverside's population-employment ratio is } \frac{7,052}{234} = 30.1$$

$$\text{Forestville's ratio is } \frac{6,758}{123} = 54.94$$

Interpretation

These data are self-explanatory. Some communities are stronger retail markets than others. As many small communities can attest, a Walmart or similar large-scale discount retailer can alter these data overnight.

Trade Capture Analysis

Trade capture analysis provides an estimate of the number of customers available for retail purchases based on proximity and population size by retail subsector. This method is determined by dividing the actual sales in the locality by the state per capita sales, adjusted by relative income.

The following information is needed:

1. Actual retail sales for a particular segment of the local economy in two periods (these data may be obtained from local sales tax information or through such sources as the *Census of Business*)
2. State per capita expenditure in the sector under examination
3. Local or county per capita income for the years being analyzed (this information may be found from a variety of sources in many states)
4. State per capita income for the years under study (this information may be obtained from the state revenue department)

Calculating Trade Area Capture

$$\text{Trade capture} = \frac{\text{Actual retail sales of merchandise type J}}{\text{State per capita expenditure for merchandise type J} \times \frac{\text{Local/county per capita income}}{\text{State per capita income}}}$$

Using *Census of Business* data, we can calculate the sales in a given area based on the size of its retail base.[2]

Retail Category	1975 Actual Sales (000s)	1975 Trade Capture (000s)	1980 Actual Sales (000s)	1980 Trade Capture (000s)
Food	5,122	14,320	9,241	17,203
Auto	3,301	10,441	9,341	17,656
Furniture	1,527	19,489	2,603	21,992
Eat and drink	2,187	12,962	2,889	10,885
Misc. retail	1,300	7,399	1,599	6,113
Total retail	22,505	13,500	37,971	15,270

1. The trade for auto sales 1975
$$\frac{3,301,000}{390.12 \times (3,457/4,266)} = \frac{3,301,000}{316.14} = 10,441$$

2. Trade captures auto sales 1980
$$\frac{9,341,000}{655.61 \times (5,539/6,864)} = \frac{9,341,000}{529.05} = 17,656$$

Interpretation

In 1975, Riverside captured the purchasing power of 10,446 persons and, by 1980, of another 7,234 persons. One caveat here: If a family of four purchased a vehicle in Riverside, they would be listed as four purchasers. Nonetheless, with certain reasonable interpretation, this information is valuable.

Pull Factor Analysis

Knowledge of the trade capture leads almost inevitably to speculating how the community can capture a larger sales volume as a means of creating more jobs. The pull factor analysis lets the community approximate the percentage of sales occurring from customers outside the immediate community boundary. This analysis presumes considerable community distinctiveness and distance between communities, as in rural areas and other places with low-density settlement patterns. With some adjustment, it can be used for the neighborhood level.

The formula for estimating the pull factor is shown in the following example: If Riverside (our hypothetical community) had a 1975 population of 6,545 with auto sales of 10,446 units, and a 1980 population of 7,065 with auto sales of 17,680 units, then:

$$\text{Pull factor} = \frac{\text{Trade area capture estimate}}{\text{Community population}}$$

$$\text{1975 Riverside auto sales} = \frac{10,446}{6,545} = 1.59$$

$$\text{1985 Riverside auto sales} = \frac{17,680}{7,065} = 2.50$$

Interpretation

It is easy to see that the number of customers purchasing cars from Riverside was half of the city population in 1975 and three times the city population in 1985. In essence, Riverside should move to protect its advantages in auto sales through community policies that maintain the dealerships.

Retail Gravity Analysis

The basic concept of a gravity analysis is to model shopper behavior by assessing the size and distance between market centers. Obviously, these are only estimates, given that some areas have considerably greater appeal for retail customers than other areas. One of the methods used to calculate the interplay between size and distance is called *Reilly's law.*

Using this technique, we can calculate the relative commercial attraction strength of certain areas. The following information is needed:

1. The population for each city being examined within the region under study (this information is usually available from the local chamber of commerce or a number of city offices)
2. The road distance between the communities under study

Calculating Reilly's Law

Trade area can be calculated via Reilly's law by determining the relationships between the two factors under study—size and distance—as:

$$\text{Extent of trade area} = \frac{\text{Distance between City X and City Y}}{1 + \sqrt{\dfrac{\text{Population of large community}}{\text{Population of smaller community}}}}$$

City	Population	Distance from Riverside
Riverside	9,000	0
Forestville	3,000	15
Miamifalls	5,000	25

Calculation for Riverside and Forestville:

$$\frac{15}{1 + \sqrt{\frac{9,000}{3,000}}} + \frac{15}{1 + \sqrt{3}} + \frac{15}{1 + 1.72} = 5.5 \text{ miles from Forestville}$$

Note that there is an alternative method developed by Stone and McConnon that factors in retail spending patterns (Deller et al. 1991).

Interpretation

The Riverside-Forestville trade area extends from Riverside up to 5.5 miles from Forestville. Clearly, Forestville can only be attractive if it offers specialty goods or a compact shopping center attractive to the customer base of the region. Size is not determinative, but it is powerful, as these results show.

Community Economic Developmental Capacity

Information on the status of the economy alone is no indicator of the ability of a community to in fact engage in economic development. Therefore, in addition to determining the economic capacity of the area, another analysis of the community institutional profile is required. This is a softer but no less important analytical process. It requires subjective judgments about the institutions and the willingness of the community to take necessary actions to alter the community's material conditions. The findings of existing research on successful local development programs strongly support the thesis that a development preparedness study is a fundamental step in local economic planning.

Successful local economic development efforts *induce* development by ensuring that the correct institutional systems are in place rather than by using gimmicks to produce or purchase economic improvements. To date, however, the development preparedness dimension of the development planning process has received far too little attention among academics, policymakers, and practitioners.

To make a serious assessment of the capacity of a community, area, or region to undertake long-term integrated socioeconomic development, information will be needed on the local institutional system, including the following aspects:

Community-Based Institutions. This includes the ability of service clubs, voluntary organizations, church organizations, social service organizations, and neighborhood or community groups to contribute to economic development programs.

Economic Structures. Organizations with a local economic focus, such as chambers of commerce, merchant associations, local development corporations, labor organizations, and state development agencies, can contribute a considerable amount of skill to the economic development process. It is essential to determine what each can contribute and to assess their economic development strength and credibility.

Political Institutions. Local government is the key to local economic development; therefore, local government bodies must have substantive capacity through boards, commissions, and agencies (redevelopment and so on) to be full participants in the process.

Financial Institutions. These include banks, building societies, venture capital groups, local development corporations, community development corporations, and small business assistance groups, which need to be mobilized to support local economic development. An assessment needs to be made of their ability to contribute.

Education and Training Institutions. Education, particularly higher education, can be a major resource for economic development. It is important to position education and training institutions to provide both skilled human resources as well as research and development expertise to the economic development program. Some determination has to be made of their willingness and capacity to become involved in this manner.

The relationships among the institutions are as important as their existence. Good working relations among them are essential because they all have a role to play in the local economic and employment development process.

To engage in economic development, a community must have the capacity to perform five development functions through its local institutions: (1) economic planning, (2) social and community resource development, (3) physical and land use planning capacity, (4) commercial and industrial targeted marketing, and (5) local finance capacity. An analysis of these items provides a framework for the possible roles and strategies that a community can or should initiate.

The first function, economic planning, as well as the last, community finance, are essentially preconditions acting as the boundaries of development planning. Many communities that possess the leadership fail to commit sufficient resources to developing the economic database or developing the fiscal capacity to develop projects. However, some communities ask for consultant reports on the local economy without sufficient local interest or will to implement them. The remaining three categories—social planning capacity, physical and land use planning capacity, and commercial and industrial targeted marketing—are the operational dimensions of local economic development.

Public officials can provide the necessary leadership and resources for physical improvements in most communities. During the manufacturing era, physical development simply implied low taxes, adequate zoning, and public infrastructure with sufficient low-cost industrial and commercial land. Today, physical development requires the implementation of aggressive policies on land use, housing, community and household services, and community beautification that complement a well-thought-out municipal economic development policy.

Commercial/industrial targeting, marketing, and financing form the core of the business development portion of any local development program. Economic development targets should be clearly selected to optimize existing human and physical resources. They should also be selected to increase the competitive advantages of the community through optimal integration of existing industries and to improve the capacity of the community to attract complementary enterprises.

Social resource development, the remaining capacity area, is the integrating component for any economic strategy plan. As stated previously, jobs need to be developed to fit the population rather than making the people fit the jobs. Furthermore, technical and higher-education resources should be focused on developing intellectual resources for the industrial base and not merely training students for export. Communities, regardless of size, need to develop meaningful development strategies after completing the expanded capacity assessment instrument. Most communities lack sufficient development capacity.

For example, many communities have little or no capacity to undertake development planning—they lack databases for formulating integrated economic and social development plans. These communities generally lack the capacity to coordinate the development process, as evidenced by the absence of a coordinating organization. Moreover, few communities have well-developed promotion programs that feature identifiable community themes, focused tourism strategies, and targeted industrial and commercial plans.

The willingness to take a calculated risk is another essential ingredient for local economic development. Communities must be willing to use their resources and take the chances necessary to alter their circumstances.

Good Information and Good Decisions

Selectivity in information gathering cannot be overemphasized. Regardless of how much planning experience a community has, care should be taken to think out data-gathering activities before beginning. The value of different data should be differentiated as essential, useful and desirable, useful but probably not essential, or of still lower priority. Long-winded reports, reams of paper, and data for the sake of data only add to the burden of planning. Other key prerequisites of good planning information include the following:

- The data collected should be comparable. Data alone may be totally meaningless unless the users can readily equate them with other data.
- Consistency, or the ability to reduce information to a "common denominator," is crucial.
- The data held must be accessible to rapid retrieval. Information buried in an antiquated filing system may be selective, comparable, and current, but it will be of no value if it cannot be retrieved when needed.

Community Participation in Information Gathering

It is important to recognize that examining the community's socioeconomic base, needs, and development capacity is much more than a process of information gathering. It measures the community's ability to take steps to develop its own resources. Most of the necessary data for this type of analysis can probably be found right next door in a local planning or service agency. A socioeconomic and development capacity assessment in which the community is approached directly, however, can be a very effective constituency-building strategy. Of course, there are limitations. If an early deadline for funding looms, there may not be time to go to the community at large. Nor is the direct approach the least expensive method of gathering information. Still, the concept of broad community involvement in a needs assessment has several major advantages.

Applying Analytical Tools to Build an Economic Development Plan

A case study is appended at the end of this chapter as an illustration of how many of the analytical tools described above are employed in the development of an economic development plan.

Summary and Conclusion

Analytical techniques alone neither help communities identify their development problems nor identify the means of solving them. Every problem presents some unique dimensions requiring insight and innovative investigative strategies. The techniques presented here are standard techniques that may prove useful in suggesting either the point of intervention or the direction. These tools should not guide the development process but assist those persons working on local economic development to devise measures of where they are, where they might go, and what impacts are likely based on the alternative proposals presented. The ultimate goals are to increase local employment, distribute wealth, and reduce community dependency. These measuring systems help us assess the degree to which the community needs to improve with respect to these dimensions, and determine, after implementing new policies, whether it has affected any or all of them.

Notes

1. Many of the examples used in this chapter have been adapted from Huestedde, Shaffer, and Pulver (1984).

2. The source of the following calculations is Huestedde, Shaffer, and Pulver (1984).

References and Suggested Reading

Agajanian, S. 1987. *California Planner.* Sacramento: California Chapter, American Planning Association.

Beed, Thomas, and Robert Stimson, eds. 1985. *Survey Interviewing: Theory and Techniques.* Sydney: Allen & Unwin.

Bendavid-Val, Avrom. 1974. *Regional Economic Analysis for Practitioners.* New York: Praeger.

Bendavid-Val, Avrom. 1980. *Local Economic Development Planning: From Goals to Projects.* Planning Advisory Service Report Number 353. Chicago: American Planning Association. (Available from the publisher, 1313 E. 60th Street, Chicago, IL)

Blakely, Edward J. 1979. *Community Development Research.* New York: Human Sciences Press.

Bosscher, Robert, and Kenneth Voytek. 1990. *Local Strategic Planning: A Primer for Local Area Analysis.* Washington, DC: U.S. Department of Commerce.

Braeschler, Curtis, John A. Kuehn, and John Croll. 1977. *The Community Economic Base: How to Compute, Evaluate and Use.* Guide DM 3005. Columbia, MO: University of Missouri Extension.

Butler, G. J., and T. D. Mandeville. 1981. *Regional Economics: An Australian Introduction.* St. Lucia, Australia: University of Queensland Press.

Deller, Steven, James C. McConnon, Jr., John Holden, and Kenneth Stone. 1991. The Measurement of a Community's Retail Market. *Journal of the Community Development Society* 22 (2): 68-83.

Gibson, Lay James, and Marshall A. Warden. 1981. Estimating the Economic Base Multiplier: A Test of Alternative Procedures. *Economic Geography* 57:146-159.

Goldstein, Benjamin, and Ross Davis, eds. 1977. *Neighborhoods in the Urban Economy: Dynamics of Decline and Revitalization.* Lexington, MA: Lexington.

Harmston, F. K. 1983. *The Community as an Economic System.* Ames, IA: Iowa State University Press.

Huestedde, Ron, R. Shaffer, and G. Pulver. 1984 (November). *Community Economic Analysis: A How-to Manual.* Ames, IA: North Central Regional Center for Rural Development.

Isard, W. 1975. *Introduction to Regional Science.* New York: Prentice Hall.

Isserman, Andrew. 1977a. A Bracketing Approach for Estimating Regional Economic Impact Multipliers and a Procedure for Assessing Their Accuracy. *Environment Planning A* 9:1003-1011.

Isserman, Andrew. 1977b. The Location Quotient Approach to Estimating Regional Economic Impacts. *Journal of the American Institute of Planners* 57:33-41.

Jensen, R. C., T. Mandeville, and N. Karunarten. 1979. *Regional Economic Planning.* London: Croom Helm.

Jensen, R. C., and G. R. West. 1983. The Nature of Australian Regional Input-Output Multipliers. *Prometheus* 1 (1): 202-221.

Jensen, R. C., and G. R. West. 1984. *A Graduated Approach to the Construction of Input-Output Tables.* Paper presented to the Input-Output Workshop of the Ninth Conference of the Regional Science Association, Australia and New Zealand Section, University of Melbourne, December 3-5.

Kruckeberg, Donald A., and Arthur L. Silvers. 1974. *Urban Planning Analysis: Methods and Models.* New York: Wiley.

Landis, John. 1986. Electronic Spreadsheets in Planning. *Journal of the American Planning Association* 51 (2, Spring), Planner's Notebook section.

Lewis, E., and R. E. Howells. 1981. *Coping with Growth: Community Needs Assessment Techniques.* Corvallis, OR: Oregon State University, Western Rural Development Center.

MacCannell, Dean. 1979. The Elementary Structure of Community. In *Community Development Research* edited by E. J. Blakely. New York: Human Sciences Press.

Mahood, S. T., and A. K. Ghosh, eds. 1979. *Handbook for Community Economic Development.* Sponsored by the U.S. Department of Commerce, Economic Development Administration, Washington, DC. Los Angeles: Community Research Group of the East Los Angeles Community Union.

Murray, James. 1978, June. Population-Employment Ratios as Supplement to Location Quotients and Threshold Estimates. *Community Economics* 21.

Rochin, Rufugio. 1986. *A Teaching Manual for Community Economic Development.* Davis, CA: University of California, Department of Agricultural Economics.

Shaffer, Ron. 1989. *Community Economics.* Ames, IA: Iowa State University Press.

Stimson, Robert, and Edward J. Blakely. 1992. Brisbane's Gateway Strategy for Economic Development: The Potential Impact of Intermodal International Transportation Initiatives. Working Paper No. 548. Berkeley, CA: Institute of Urban and Regional Development, University of California.

Voytek, Kenneth, and Harold Wolman. 1990. *Local Strategic Planning: A Manual for Local Economic Analysis.* Washington, DC: U.S. Dept. of Commerce, Economic Development Administration.

Young, Frank, and Ruth Young. 1973. *Comparative Studies of Community Growth.* Monograph No. 2, Rural Sociological Society. Morgantown, WV: West Virginia University Press.

Wadsworth, Y. n.d. *Do It Yourself Social Research.* Victorian Council of Social Service, mimeo. Melbourne, Australia: Melbourne Family Care Organization.

APPENDIX 5A

Merging Strategic Planning and Analysis to Build an Economic Development Plan

The first edition of this book did not address how data and attitudes help shape a strategic economic development plan. The case below illustrates the integration of strategic planning with quantitative analysis in the evolution of an economic development plan. It incorporates a number of the tools discussed in Chapters 4 and 5. This is an actual case that has been slightly abridged to fit this book. Those interested in the entire case and the data associated with it can write to me.

The Brisbane Case

Brisbane city is at the center of a metropolis with a rapidly growing population. What happens in the region has direct a impact on the city's economic activities, traffic, environment, and general quality of life.

The Brisbane Plan was an economic development strategy that involved detailed research into a wide range of aspects of the city's internal trends and its regional environment.

Demographic Trends

Brisbane city could become the "hole in the doughnut" of the metropolis. The city's population is aging. Young couple households are leaving for cheaper land and housing packages available in the jurisdictions of surrounding suburban local governments. Brisbane urban population can be expected to decline within the next decade or so on the basis of current trends and policies. Its population age structure will become increasingly "unbalanced," that is, underrepresented by young families and overrepresented by middle-aged and older people and young single adults. Well over half the households in the city are now one- or two-person households (see Table 5A.1).

Meanwhile, it is projected that the surrounding suburban areas will grow rapidly, with the Brisbane metropolitan area population likely to reach 1,753,000 by 2011.

Scenarios were developed to look at how the city of Brisbane might address its current trend towards future population decline and the increasingly "unbalanced" population structure. These scenarios indicate that if appropriate policies are implemented to cater to the first-time home buyer households, then longer-term population growth could be assured, resulting in the achievement of a more "balanced" population structure. The city's population could increase to about 850,000 by 2011.

Economic Trends

The Brisbane regional economy is projected to grow, particularly in such service industries as tourism, recreation, and personal services. Industries based on new technology are assuming a greater role in total economic activity and are growing at rates faster than the economy as a whole. The labor force is growing fastest in the service, public administration, and retail/wholesale trade areas; female participation is also growing rapidly.

It is projected that there will be a change in the demand for industrial land from the inner to the outer suburbs, and an increase in demand for high-quality industrial land for technology-based enterprises now in the inner city and in close proximity to client firms in the services sector. Demand for industrial land near the airport and seaports areas, particularly for transport-related activities, will increase.

In retailing there will be a continuing suburbanization of retail trading, but on the basis of existing population trends there will be little need for development of new regional or subregional shopping centers in Brisbane city. Trends in retailing indicate demand for freestanding warehouses and for facilities oriented towards leisure and recreation within existing centers.

For offices there will be strong growth in floor-space demand in the Central Business District and its fringe areas, with suburban office demand also increasing.

Social Trends

Brisbane city's aging population and changing structure of households will need increased services for elderly people, single-parent female-head households, and women in the labor force. There is also increasing ethnic diversity in Brisbane city, with increasing immigration from Asia. Demand will increase for more diverse recreation, sports, and cultural services and facilities.

Many communities are changing rapidly. "Gentrification" is beginning in the inner suburbs. There is an increasing degree of mismatch between the location of services and where the people they are intended to serve live.

Methodology

The methodology used for the economic development strategy is based on the use of sophisticated input-output tools, along with qualitative measures. It estimated the economic implications of a number of strategic development options and opportunities for Brisbane city and the region. This methodology consisted of two stages.

Stage 1: a projection component, estimating the Brisbane regional economy in a "hands-off" development scenario; that is, the path the regional economy is expected to follow in the absence of major strategic planning initiatives

Stage 2: an impact component, estimating the expected impact of a number of alternative strategic planning scenarios on the "hands-off" economy, or the changes in economic indicators that would accompany the implementation of these scenarios

The projection stage of this strategic planning exercise is described below. Refer to Figure 5A.1 for an overview.

The Projection Methodology

This section describes the procedures used to develop projections of the Brisbane regional economy for the years 2001 and 2011, which were the reference years for the study. It provides the summary results at the 19-sector level. The procedures used modified those developed for a recent analysis of a similar area—the

Table 5A.1 Household Types for Brisbane City and Balance Area of the Brisbane Statistical Division (BSD), 1986

Household Types	Brisbane City		Rest of BSD	
	Households (000s)	%	Households (000s)	%
Parents with dependents	9,814	3.91	6,727	4.91
Parent, adult family members, and dependents	3,920	1.56	2,367	1.73
Couple	57,763	23.00	29,473	21.52
Couple with dependents	54,488	21.70	51,719	37.76
Couple and adult family members	22,767	9.07	10,348	7.56
Couple, adult family members, and dependents	17,073	6.80	11,493	8.39
Related adults	16,016	6.38	5,039	3.68
Lone person	55,489	22.10	16,266	11.88
Group	13,765	5.48	3,521	2.57
Total households	251,095	100.00	136,953	100.00

SOURCE: Australian Bureau of Statistics (ABS) 1986 Census.

Gladstone region. The Gladstone study combined conventional shift-share analysis and input-output tables to provide estimates of structural change in the Gladstone regional economy over the period 1974/75 to 1985/86.

The shift-share component enabled growth in industry gross output to be attributed to national growth factors, industry-mix factors, and region-specific or differential factors. Extensions to shift-share analysis and the use of input-output tables enabled the estimation of these effects within each cell of the input-output table, thus allowing a more detailed study of the effects of regional economic structural change.

The Gladstone study extended the study of structural change to both a technology effect, incorporating change in cells within each industry, or changes in input structure, and to changes in regional import patterns. This study used a similar technique.

Three phases were identified in the construction of the input-output tables showing the structure of the Brisbane regional economy in the years 2001 and 2011. These were:

1. The identification and measurement of the components of structural change in the Brisbane-Moreton regional economy between the years 1985/86 and 1989/90. This was used as a possible source of data for projection of the regional economy to the two reference years 2001 and 2011.
2. The construction of a 1989/90 input-output table of the Brisbane region, consistent with the existing 1985/86 table. The 1989/90 input-output table was used as the base table for projection of the regional economy to the reference years in Phase 3 below.
3. The projection of each component of structural change to obtain projections of the expected size and structure of the Brisbane regional economy in the two reference years.

Phase 1

This phase of the study identified and estimated the components of economic structural change in the region over the period 1985/86 to 1989/90. These estimates are reflected in Phases 2 and 3 below.

The general approach to identifying past structural change was modified from the Gladstone study and is represented in Figure 5A.1. In Phase 1, the base year table was the 1985/86 interindustry table of the region. The components of structural change refer to the change estimated to have occurred between 1986 and 1990. In Phase 3, the base year interindustry table for 2001 was the 1990 table, and for 2011 it was the projected 2001 interindustry table. The components of structural change reflect the growth assumptions of the model.

In both Phase 1 and Phase 3, each component had a similar meaning: Phase 1 attempted to estimate the contribution of that component to changes in economic structural change in the Brisbane-Moreton region in the period 1985/86 to 1989/90. Phase 3 showed the expected contribution of each component to future regional economic change in the two reference years 2001 and 2011.

In the interests of reasonable brevity, and to avoid the presentation of a large number of whole and partial interindustry tables, the empirical component of Phase 1 is not described in detail. A brief description of the four components is given below. A description of each component in the context of the projection process is given in the discussion of Phase 3. The detailed account of the methodology and the resultant input-output are not included here.

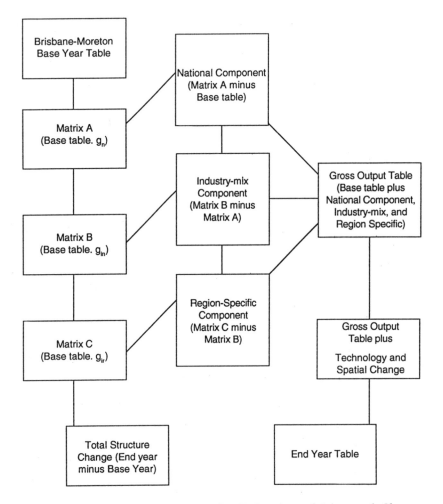

Figure 5A.1. Analytical Structure for Estimation of Structural Change 1985/86 to 1989/90, and Projection to Years 2001 and 2011, Brisbane-Moreton Economy

National Component

The national component represents that part of the transactions change, either in total or in each cell of the table, attributable to the growth of the national economy as a whole. It is a recognition of the fact that some of the growth that has occurred in the Brisbane regional economy has occurred because of the economic environment of the national economy, that is, changes in national standard of living, balance of payments, monetary and fiscal policy, and so on. Economic factors affecting the national and state economies will substantially influence the performance of the Brisbane-Moreton economy. Although each region of a nation reacts differently to the national economy as a whole, there is a sense in which each region in a national environment will exhibit characteristics of the national economy simply because it is part of that economy.

The national component was calculated according to the estimated growth rate of the gross national output over the period 1985/86 to 1989/90. The national growth factor (gn) was estimated at 1.099, which represents an average growth rate of 2.4% per annum. Matrix A, calculated by applying the factor gn to the base year table, shows the transactions that would have occurred in each cell if gross output levels of each industry had grown at the same rate as the national economy. Subtraction of the base-year table from the A matrix produced the national-growth component matrix.

The Industry-Mix Component

The industry-mix component measures the effect of the composition of industry in the Brisbane-Moreton region, due to the fact that, nationwide, some sectors grow more rapidly than others. A region specializing in slow-growth sectors will show a negative industry-mix component, whereas a region favored by a high share of rapid-growth industries will show a positive industry-mix effect.

Matrix B estimated the transactions that would have occurred if each industry at the regional level had reacted in the same way to economic circumstances in the period 1985/86 to 1989/90 as that industry at the national level. Matrix B was derived by the scaling of each industry by the estimated national growth factor for that industry $(g_{i,n})$. The subtraction of Matrix A from Matrix B produced the industry-mix effect. Brisbane region contained 8 faster-growing and 11 slower-growing industries in terms of national growth rates. The overall industry-mix effect was positive during the period 1985/86 to 1989/90, suggesting that those industries in the region with a faster national growth rate were more important in terms of output than those growing more slowly in national terms.

Region-Specific Component

The region-specific effect arises from the fact that industries grow at different rates in different regions. This occurs because some regions have a comparative advantage over others for the growth of individual industries. The region-specific effect refers to the extent to which individual industries in the Brisbane-Moreton region have grown faster or slower than the same industries at the national level because of the region's economic environment. Matrix C was estimated by the scaling up or down of each sector by the regional growth factor for that sector ($g_{i,r}$). The region-specific component is the difference between Matrix B and Matrix C. This reflects the differences between the growth rates of industries at the national and regional levels.

In the period 1985/86 to 1989/90, 8 industries in the Brisbane-Moreton region were estimated to have grown faster than their national counterparts, and 11 more slowly. The region was shown to have a comparative advantage as a service center and for some manufacturing activities.

Technology Component

The technology component reflects a number of influences—namely, changes in technology of production, input substitution, and importing patterns. It reflects the changes that occur within each column of the interindustry table, or the changes in the process of production and local input purchases. The technology component was estimated either by the identification of changes in column structure since 1985/86 or the relative change in each cell of the table.

Phase 2

The second phase in the study was the preparation of a 1989/90 interindustry table for the Brisbane-Moreton economy, to be used as the base table for projections of a future regional economic structure. The base table was constructed to represent the current state of industry in the region rather than as a "snapshot" of the economy in 1990. As current data were not readily available, the table was constructed using expected growth rates of industry based on the identified components of structural change estimated in the first phase. These estimates were validated from industry sources, where possible.

Phase 3

The final phase involved the derivation of interindustry tables showing the expected size and structure of the Brisbane-Moreton economy in the reference years 2001 and 2011. This involved the projection of each component estimated in Phase 1, and the layering of each of these components on the (now) base year 1989/90, to obtain the respective projected tables of the Brisbane region for the years 2001 and 2011.

Deriving the national and regional industry growth rates to incorporate them into the projection model was an important and complex issue. The various growth factors (g_n, $g_{i,n}$, and $g_{i,r}$) developed in Phase 1 of this study could not be applied uncritically to project the Brisbane-Moreton economy for the 11-year and 21-year periods involved in this study. The growth rates derived in Phase 1 were accepted as a basis for future growth rates, but were modified in terms of current and expected future economic circumstances and expected developments in technology and consumption patterns. These included population projections, forecast trends, and industry and professional opinions. The projected rates of change incorporated into the projection model are conservative but attempt to incorporate available opinions on likely market developments.

A brief description of the derivation of each of the components built into the projection model is now provided.

National Growth Factor

The national growth factor assumed for the period 1989/90 to 2000/2001 was 1.256, representing an average annual growth rate of the national economy of 2.1% over 11 years. The growth factor assumed for the period 2001 to 2011 was 1.195, or an average annual growth rate of 1.8%. As in the Phase 1 calculation, the national component was expected to be a substantial, and possibly dominant, factor for the Brisbane-Moreton economy, and for all sectors within the economy.

Industry-Mix Factor

The industry-mix trends identified in Phase 1 were expected to continue from 1990 to 2001 and 2001 to 2011, leading to a positive industry-mix contribution to the growth of the Brisbane-Moreton region in the two reference periods. Generally, the service sectors of the economy were expected to provide the greatest positive industry-mix effects in the future, with some manufacturing and primary industries growing significantly less than the national growth factor.

Region-Specific Factor

The relative size of the region-specific effect, in terms of the difference between the assumed growth rate of each industry at the national and regional levels, is given for the two time periods. A positive entry in these columns reflects the assumption that the industry identified will grow more rapidly at the regional level than at the national level. If these trends continue, the Brisbane-Moreton region will have a locational advantage for most manufacturing industries and for most of the service sectors, and show lower than national growth rates for the primary sectors.

Technology Factor

Structural changes within the columns of the interindustry table reflect changes in the purchasing patterns of the industries of the Brisbane-Moreton region. These changes reflect the assumption that the industries within the region will purchase, for example, more of their inputs from within the region as the diversity of the regional economy is increased. In particular, it is expected that relatively increasing purchases from the finance sector will occur, as well as from the transport, communication, and personal services sectors.

The 2001 and 2011 Interindustry Tables

From these projected estimates, interindustry tables for the Brisbane-Moreton region were derived for the years 2001 and 2011. Given that change in the level and mix of economic activity is inevitable in any economic system, it is possible to estimate the expected changes in the nature of the Brisbane-Moreton regional economy in the future. The projected interindustry tables for the years 2001 and 2011 provide several types of summary information as to the nature of the future regional economy. For example, it is possible to calculate changes in the expected size of regional economic indicators and the future contribution of each regional industry based on these indicators.

Economic Development Choices

Basically, the metropolitan area of Brisbane faces three economic development choices. These are the following:

1. Pursuing a noninterventionist role in which current policies continue, as do market forces, to influence how individual industrial sectors develop. Such a policy may be disadvantageous. There will still be economic growth resulting from the normal market forces in an expanding national and regional economy. In current trends, however, it is unlikely that this expansion will provide sufficient employment opportunities for the greatly increased workforce. It is more likely that, for a number of reasons, population increase will be accompanied by local structural unemployment unless an active strategy is undertaken to increase the economic activity in the region faster than the rate of population growth. Without such positive steps, social dislocation and severe welfare problems will result from the shortfall in employment.

Even if sufficient employment opportunities are created, this strategy will also need to ensure that sufficient additional output is generated to increase per capita gross regional product. If regional economic activity fails to achieve greater output per capita, then it can be expected that the standard of living measured by real per capita income will deteriorate.

2. Concentrating on the development of a few key industries by attracting new enterprises and encouraging development in selected existing industries, with the aim of seeding economic growth through the flow-on effects. Because of the narrow focus on a limited range of the industrial sector and the emphasis on attracting new industries, the success of this option is vulnerable to economic downturn in the national economy, international competition, and the reluctance of new business to relocate to Brisbane. This option would also require the provision of large sums of capital funding, with a high risk of failure if new entrants are not attracted to Brisbane.

3. Concentrating on broadening and deepening the existing economic structure by removing barriers to new firm entry and promoting a climate conducive to industrial restructuring, thus emphasizing the current strengths and natural advantages of the regional economy.

A broader-based economy would be better able to withstand fluctuations in the national and international economies, and as the economy developed, new industries would find it easier to relocate to the Brisbane area. Although this option would require capital outlay to provide the climate for expansion, the amount would not be as great as under the previous option, and the risk of failure is lower.

Assessing Community Industrial Preferences

In order to determine community reactions to various future industrial choices, the business and professional communities in Brisbane were surveyed regarding their reactions to various industrial sectors and to assess their potentials for development. These took two forms:

1. The holding of a series of panel discussions with a number of industry-sector leaders—drawn from the business and commerce, manufacturing, research and development, tourism, and administration sectors of the city and the region—to identify strengths, weaknesses, options, and opportunities for the regional economy and its industry sectors.

2. The collection of data through a questionnaire survey of leaders from these fields. They were asked to make judgments based on their assessments of local resources and the potential of 17 industrial sectors, using 5-point scales to assess the current capacity and future capability of the region according to 5 factors, namely *finance,_ political* and *headquarters potential, positive environment*, and *enhancing the area's international image and impact*.

These data were used to develop a matrix of the industries preferred by the community. The matrix shows the perceived strengths or weaknesses of each sector, from the perspective of community leaders. The results are summarized in Figure 5A.2.

The data revealed the following indications for the Brisbane region:

1. The old manufacturing-of-raw-materials base is seen as a dominant economic resource.
2. There are no real barriers to financing most of the raw-materials-based industries as well as transport and communications.
3. There is less financial and political capacity in the metals area.
4. There is more capacity to support past industries than resources to develop future firms.
5. Environment and image are the most significant factors with respect to community reactions, whereas the food, communications, community services, transportation, recreation (tourism), and finance sectors have the highest ratings.

Based on the survey results, community preferences indicated that the tourism/recreation, financial management, construction, environmental, and food-manufacturing industries have the strongest support. This industrial grouping could lead to economic

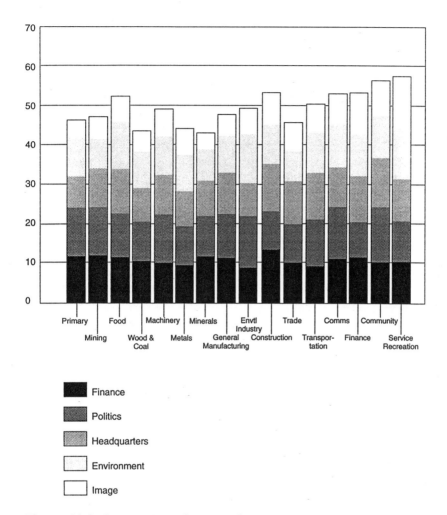

Figure 5A.2. Community Industry Preferences

development scenarios emphasizing regional services, tourism/ education, and/or transportation/distribution industries.

Scenario Building

Using the analysis of the input-output tables generated, as described earlier, it was possible to show scenarios outlining potential economic development options for the Brisbane region. The structural analysis of the region's economy identified strengths and weaknesses and the assessments of sectoral capacity and potential. These scenarios were developed to illustrate what might be plausible alternatives and were not proposed as final discrete strategies. A number of potential economic development scenarios were proposed. They took advantage of Brisbane's strategic location, its role as the state capital, its historical development as the major service center for the rapidly growing Southeast Queensland urban region, and the fact that Brisbane is well located to provide links between Australia and the Asia-Pacific region combined with community preferences.

Three important assumptions were made in developing scenarios:

1. Forecasts of regional population growth from 1.81 million in 1990 to 2.21 million in 2001 and 2.6 million in 2011
2. A regional focus for the economy
3. Quality lifestyle (weather, housing, recreation, culture, physical environment, etc.)

The scenarios developed were not mutually exclusive; rather, they tended to complement each other. However, although they were not additive in their effects, they did have a synergy effect. The five scenarios developed are summarized below.

A. Development of an international intermodal transportation center (SuperPort). This scenario capitalizes on Brisbane's strategic position in relation to the state, the rest of Australia, and the Asia-Pacific region. The city has an unusual strategic concentration at the mouth of the Brisbane River, with a seaport, an airport, and a well-developed road network and rail system, all of which have spare capacity, together with large areas of relatively underutilized land. The scenario combines these advantages to develop a modern multimodal transportation facility with the capacity to manage air and surface transport to the Asian-Pacific region.

B. Development of a technology- and education-based sector. This scenario proposes the expansion of the emerging education-, technology-, and knowledge-based industry in Brisbane. It would use the resources of the Technology Quadrangle of the four higher-education institutions and the existing research and development base in the region by developing innovation centers and related technology infrastructure. It would build on existing technology rather than relying on attracting new firms to the city or region.

There would be an emphasis on promoting and developing technologies related to the existing areas of strength and applying technology to existing natural resources.

C. Continued development of Brisbane as a regional service center. This scenario assumes that continued development of the central city area's service and administrative sectors will be required to support the growth of Brisbane and the region's population, as well as the expansion of the city's manufacturing and service base. In this scenario, Brisbane's role as a regional government center is the basis for economic growth. Public service, cultural, and community services (sports, conventions, media) would be based in Brisbane. Brisbane's share of the regional population would decrease substantially, but its headquarters role would be enhanced.

D. Brisbane as an events and destination center. The central idea of this scenario is to build on Brisbane's natural features and its image as a city attractive to residents and visitors. The scenario would promote Brisbane as the hub of travel in the region and a major destination for conventions. The city would also become the headquarters for the regional tourism industry, providing the management and supporting infrastructure. A concerted program to attract major events is an important element of this scenario.

E. Greater emphasis on adding value to the state's natural resource output. In this scenario, Brisbane and the region would strengthen their capacity to process or manufacture the raw materials mined or grown in the state. This scenario depends on the region's growth as the primary driving force for the expansion of the manufacturing base. As the regional manufacturing sector expands its capacity to add value to raw materials to meet regional markets, the potential to export to the rest of the nation and overseas will expand.

Projecting Scenario Impacts

Earlier in this chapter I described the projection methodology used to derive the interindustry tables for the region for the base year 1989/90 and for the reference years 2001 and 2011. I described the choice of development scenarios for consideration in the planning process. These scenarios were applied to determine the expected economic implications for the region if each scenario was to be successfully implemented. They appeared as changes to final demand in the projected interindustry tables. This section describes the results of this analysis.

The five scenarios were subjected to the same process of input-output analysis as the Brisbane regional economy. This was in order to assess economic impact by evaluating likely changes to gross regional output and employment levels in the years 2001 and 2011 (see Table 5A.2). In carrying out this analysis, the industry growth rates were adjusted to take into account the likely variations in industry activity levels that would result from the implementation of each scenario as part of a strategic economic plan.

At this point in the strategy's development, it is appropriate to treat each scenario as a stand-alone option, both to assess the impact and to consider the strengths and weaknesses of each scenario.

Examination of these results reveals that, in the year 2011, Scenario A (development of an international intermodal transportation center) offers the greatest opportunity both for employment and additional gross regional product. Scenario B (expansion of the region's technology- and education-based sector) ranks second, followed by Scenario D (Brisbane as an events and destination center). Scenarios C and E were similar in creation of employment opportunities, although their ranking reversed for the creation of additional gross product.

By the year 2011, Scenario B would overtake Scenario A in additional regional output, although still ranked second in employment opportunities. Scenario D remains ranked third.

Comparisons between the five scenarios and the interindustry data, taking into account the additional household income generated both in total and per additional employment opportunity, suggest a few conclusions:

1. Early expansion of the transportation-related industries provides rapid shorter-term growth, overtaken eventually by the technology and education scenario. This suggests that technology-sector capital investments and the emphasis on human resource development provide a longer lead time for income generation than does investment in the transportation sector but offer greater long-term potential for growth.

2. The services scenario provides lower growth potential than Brisbane's historical development as a service center would suggest, probably because the sector is already well developed in the region and can cope with the rapid population growth and demand for services without a proportional growth in employment; that is, higher productivity and output levels can be achieved.

Table 5A.2 Economic Impacts–Development Scenarios Expected Additional Contributions

	Scenario				
	A	B	C	D	E
2001					
GRP ($ million)	6,873	4,055	3,057	3,601	2,711
Household income ($ million)	1,876	1,427	618	1,026	636
Employment (000s)	78.8	58.9	24.4	45.8	28.4
Household income/ additional employee ($1,000s)	23.8	24.2	25.3	22.4	22.4
2011					
GRP ($ million)	12,987	13,521	6,303	7,594	5,504
Household income ($ million)	3,494	3,334	1,372	2,006	1,292
Employment (000s)	135.9	128.2	51.8	80.4	52.0
Household income/ additional employee ($1,000s)	25.7	26.0	26.5	25.0	24.8

3. Lower per capita household income generation in Scenarios A, D, and E reflects the structure of those industry sectors, with high capital-to-labor ratios, lower wage structures with less dependence on professional or skilled workers, and probably an existing industry structure that can expand to meet market demand by addition of lower-paid workers without requiring similar growth at management or technical levels.

4. Although the scenario of adding value by transformation of the mining and rural outputs offers good potential for growth, there may be real limits on expansion of the manufacturing sector imposed by the small local markets (although these will grow), by the limitations of the international market, and by Brisbane's historical lack of an extensive manufacturing base.

Gateway: An Integrated Strategy

The Economic Development Strategy may best be described as a Gateway Strategy, which makes use of Brisbane's key position as a link between the emerging economic strengths of the Asia-Pacific region. The city is the main point of entry to the fastest growing region in the nation, and its advantageous position makes it a transportation hub for passenger and cargo services.

Although an industry or sector is often isolated for the purposes of evaluation, it is not possible to quarantine Brisbane from the regional economy. Separation of particular sectors or industries does not imply that a sector or industry can be subjected to actions or policies that will not affect the wider economy.

Any region's economy is an integrated network of industries; increased activity in any one industry or sector is reflected through flow-on effects (the economic multiplier) throughout the economy. These flow-ons result from increased demand for inputs into the expanded industry (for example, labor, capital, raw materials or components, transport, business services) and from the increased outputs that generate wealth through the distribution of profits, dividends, and wages.

The Gateway Strategy provides a basis for realizing Brisbane's economic potential in the context of the totally integrated regional economy. It utilizes and develops the quality of the city's human resources in service industries and technology, integrating the city's economic development with that of the region. It ensures that the industry location within the region builds on the strengths of the regional municipalities, preserving land for future development use, protecting economic resources, and ensuring compatibility of development between adjoining land uses. It encourages the expansion of existing industries and the development of new environmentally sensitive, energy- and resource-efficient industries compatible with social and residential development. Maximum advantage is taken of Brisbane's strategic location as the transportation focus for the region and the capacity of its transportation infrastructure. That advantage increases Brisbane's potential for headquarters functions and services industries by facilitating modern office development and communications capability in the Central Business District and the designated regional business centers.

The strategy identifies and maximizes the most appropriate and likely components of each of the five selected industrial sectors and builds on their synergy through the flow-on effects created by that activity in one sector. In essence, the city can take advantage and shape the direction of economic activity within the region in a way that will contribute to the simultaneous economic improvement of the city and regional economies. Figure 5A.3 illustrates the employment opportunity effects of the Gateway Strategy.

Every community is faced with strategic choices. These choices can be based on mere guesses and hunches or impressions of the potential of the community's economy or rely on sophisticated econometrics models. Sometimes a good guess or hunch is as valuable as all the analysis in the world. For example, Walt Disney built one of the world's most well-known entertainment industries in a community—Anaheim—that seemed at variance with all of the existing data. On the other hand, good guesses are sometimes pain-fully wrong as many would-be high-tech parks can easily attest. The best approach is to balance the community's intuitions with real data and analytical techniques. The case presented in this appendix shows how the community's interest and intuitions can be assessed against their socioeconomic goals.

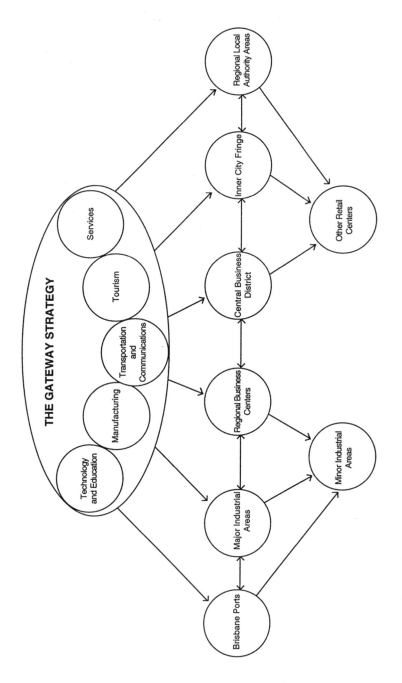

Figure 5A.3. Employment Opportunity Effects

132

6

Selecting a Local Economic Development Strategy

Strategic planning is the new "buzz word" in the planning profession. Selecting a strategy and announcing it publicly has occupied much of the attention of planning bureaucracies in many of the nation's cities. Strategic planning has become popular, in part, because of turn-of-the-century mentality. As the 20th century draws to a close, the shaping of the next century has become a national preoccupation. At the end of the 19th century, similar events took place. The Great Chicago Exhibition of 1893 focused on foretelling the developments of the 20th century. Interestingly enough, that exhibition placed planning for a modern America at its core. As this century ends, planning is again a major priority. However, much of today's planning appears to be a rejection of the industrial model forecast as the national ideal only 100 years ago. Whatever direction planning takes, interest has definitely increased in how planning tools must be used to guide the nation. The Clinton administration has pledged to embark on a large-scale search for national consensus on economic directions. As a result, there will be increasing emphasis at the state and city levels to engage in strategic economic planning. This chapter is based on how economic development planning fits with the overall strategic objectives of a community and the nation. *Strategies*, in this context, are planned actions for specific development goals of a community based on the options and opportunities available.

Projects emerge from specific courses of action undertaken within a given strategy. The distinction between strategies and projects is necessary because, in too many instances, a single project or group of projects with no particular focus are often described as a "local economic strategy." In most instances, these ad hoc efforts are reactions to current circumstances. The most common rationale for the majority of these uncoordinated efforts is that "something had to be done."

Strategies, however, constitute an overarching set of principles that form concepts to guide general to specific actions (i.e., from goal to strategies to projects). Strategies are valuable in forcing clarity of thought and generating consensus during the local economic development process. It is, therefore, important to examine alternative strategies along with a set of specific projects or proposals as the basic building blocks of an economic development plan. This is the approach that has been adopted here. Roger Kemp's *Strategic Planning in Local Government* (1992) provides a wide selection of informative strategic planning cases that should be examined by students and/or practitioners before embarking on strategic planning projects.

The Goals of
Local Economic Development

The context for any economic development strategy is to do the following:

Build quality jobs for the current population. The thrust of economic and employment planning is to build employment for and with the resident population of the community. This is preferable to relying on approaches that attract new employers who may require a different set of skills than those possessed by or that can be developed with the resident labor pool.

Achieve local economic stability. Economic development will be successful only if the community has a specific approach to meet all the needs of business (i.e., land, finance, labor, infrastructure, and technical assistance in addition to labor). Many cities don't even know the locations of available industrial sites or how a firm can identify requisite financing. If a community wants to obtain and retain jobs, it must have all its economic resources and socioeconomic data available in an organized form.

Build a diverse economic and employment base. No community with a single employer or set of employers is safe from fluctuating employment. Regardless of whether the community is high-tech or low-tech, it must have a broad base to provide continuing employment opportunities for residents.

Prerequisites for
Successful Strategy Formulation

The essential starting points for designing appropriate development strategies are the socioeconomic base analysis and development capacity analysis described in the previous chapter. At the conclusion of these analyses, a community can identify its opportunities, its challenges, and the resources available to meet them. A community should also, at this point, state explicitly its economic development goals. These goals should specify those sectors and groups that economic development is to serve.

When determining target sectors, industries, and social groups, one should keep several important points in mind. The best job-creation strategy is one that stresses increasing "basic" employment. Basic employment entails business activities that provide services primarily outside the local area via the sale of goods and services but whose revenue is directed to the local area in the form of wages, payments to local suppliers, and capital expenditures. Examples of business activities that could represent basic employment include resource-based manufacturing, transportation, and wholesale industrial sectors. When an office industry represents a headquarters for services, the result also is an increase in basic employment. "Nonbasic" employment, however, is associated with services and business activity that primarily serves the local area; revenue sources, therefore, are from the community. As a result, an analysis of the economic base should consider the following:

- Determining which sectors play a dominant role in the local economy in terms of jobs, sales, taxes paid, and linkages to other local industries—for example, agriculture, forestry, some types of manufacturing, commerce, services, and government
- Identifying important linkages between the local economy and the external economy in order to gauge the extent to which local sectors respond to changes in the regional, national, and international economy
- Assessing the local potential for economic growth, stability, and decline, and identifying contingencies that would initiate or complement each trend
- Identifying contingencies important to the local population or political leadership that could have major impacts on jobs, sales, incomes, public revenues and expenditures, economic productivity, job quality, and local quality of life

With target economic sectors and social groups as firm points of reference, the process of identifying and assessing alternative economic development strategies will be a clear but difficult task. The identified economic development opportunities probably will not provide all of the employment for the disadvantaged groups of the community. More than one economic development strategy will be required to meet local needs.

Selecting Strategic Options

It is well known that localities with severe economic problems face a morass of complex, interrelated issues. That is, such local situations result from a mixture of people needing jobs, firms closing or leaving, and factors that attract or produce more jobs. No single facet of the economic/employment problem is easily resolved. Too frequently, however, a single dimension or aspect of the problem is highlighted without full comprehension of its interaction with other components. In order to provide a better perspective of the situation faced by local governments and community groups, the analytical model created here portrays options as the components of an overall local development strategy. Each of the strategic options listed presents an alternative approach to meeting one or more aspects of a community's needs.

City and neighborhood leaders must decide how each of these components should be combined in order to create a unique strategy that will secure the types of jobs and requisite job balance for the community, given the resources available and local development objectives.

There are four strategic approaches. These are strategies that emphasize (1) locality or physical development strategy, (2) business development, (3) human resources development, and (4) community-based development. In most instances, a strategic plan will incorporate different elements of these approaches depending on local need and circumstances.

Each of these major components is part of the mixture of approaches the community develops to create a local economic development strategy. That is, strategies, like problems, are unique in their development and in their application.

The Locality Development Strategy Option
(The Built Environment Dimension)

As major developers of parks, roads, drainage, parking lots, and, in some cases, water and electricity supply, local governments influence the establishment and operating costs of businesses. By developing a program to upgrade a locality designated for industrial and/or commercial uses, local government may have a positive influence on local business development. If there are already a number of natural and community service advantages, a community group might also decide to upgrade them because potential entrepreneurs and managers often take into consideration amenities that raise the quality of life.

The tools for accomplishing locality development goals are numerous and include the following:

- *Planning and development controls.* Positive use of these controls improves the image of councils with business and has a positive influence on the investment climate.
- *Townscaping.* Economic upturn in town commercial centers can be achieved by making improvements to the street (e.g., planting shade trees) and to local business premises (e.g., improving window displays and the physical standards of commercial buildings).
- *Household services and housing.* A well-housed and -serviced labor force is an inducement to businesses; in addition, activities in this sector also have the potential to generate employment.

The full array of locality development tools available is discussed in Chapter 7.

The Business Development Strategy Option
(The Demand Side)

In many places, there are not enough jobs within existing firms to meet local population needs. Therefore, new ways have to be found to encourage new businesses, to attract existing businesses to relocate in the area, and to sustain and expand existing local firms. This should result in a net increase in total jobs. Several mechanisms can be employed, including the following:

- *Small business assistance centers* to provide accessible management training, counseling, consulting, and research services for small firms as a means of improving their economic performance and possibly helping them to expand their workforce

- *Technology and business parks* to provide the specific infrastructure requirements of sought-after industries
- *Venture financing companies* to provide venture capital to selected firms unable to obtain financing from other traditional lending institutions
- *One-stop business information centers* to expedite the information needs of existing and potential new businesses

The methods used to improve local business development are discussed in greater detail in Chapter 8.

The Human Resource Development Option (The Supply Side)

This strategic option forges close connections between the employment needs of certain segments of the local population and the job-formation process. The goal is to alter the human resource system in ways that increase opportunities for good jobs for the unemployed and underemployed in the community. The methods used include the following:

- *Customized training,* providing the employer with specific training based on the firm's requirements
- *Targeted placement,* ensuring that employers who receive government assistance are obliged to hire qualified local personnel as the first source of employees
- *Local employment program,* developing employment offices in the community that run training and personal skills development programs to help especially disadvantaged social groups gain employment or acquire increased skills

These approaches are coupled with several related tools that can be used to develop local human resources, as discussed in Chapter 9.

The Community-Based Employment Development Strategy Option (The Neighborhood Dimension)

This option is designed to promote economic development at the neighborhood/small community level and create employment opportunities for persons who are long-term unemployed, youth, or groups seeking to play unconventional roles in the economic system.

In this approach, activities are intended to function as intermediaries between the social welfare system and the local economy.
They are aimed at providing alternative employment opportunities for individuals who require new skills or skill upgrading. In
addition, these alternative socioeconomic structures may serve
those who wish to contribute to society through enterprise firms
that promote economic democracy. The basic activities include the
following:

- *Community-based development organizations:* nonprofit organizations that own and/or operate entrepreneurial activities and also provide a wide range of community services (the purpose of providing
  both business and services in the same organization is to facilitate,
  without stress, the movement of usually difficult-to-employ individuals through the progressive stages of employment)
- *Cooperatives:* worker-owned and -managed businesses in which the
  group shares the responsibility and liability for generating wealth
  and/or employment using jointly held resources

Clearly, these alternative approaches to local economic development can and will be mixed to meet the requirements of a given
situation. Different strategy options will suit different socioeconomic circumstances. Each of the three basic types of socioeconom-
ies—growing, unstable, and those that are restructuring, including
distressed mature economies and declining or chronically depressed
economies—requires different strategies or approaches. These are
described in Chapter 10.

Common Traps in Strategy Formulation

When selecting economic development strategies, community
leaders will face several traps. These pitfalls are usually derived
from civic leaders' anxiety to move quickly and get results. In their
haste, however, they frequently overlook important fundamentals,
such as the following:

1. *Depending too much on government programs.* Local development practitioners often accept without question national government grants, loans, or other programs such as enterprise zones
or similar approaches. Public officials usually attempt to make local
needs fit national government programs rather than basing approaches on local needs. All too often, such attempts overlook

employment or economic problems and divert attention from the real assets of the community and/or its economic development limitations. Moreover, when the government program goes away, often the local economic strategy also disappears.

2. *Letting the tool(s) determine the strategy.* Civic leaders sometimes confuse a particular development tool with a comprehensive strategy plan. Instruments such as industrial parks, small business assistance, one-stop business information offices, tax relief, or even more sophisticated public-private financial schemes are ingredients of an economic strategy, not the plan itself.

3. *Starting at the wrong end of the problem.* Although employment creation is the usual goal of most communities, human resources are seldom the focus of local economic planning. Usually, local planning groups attempt to attract any firm in the anticipation that the local workforce will obtain most of the jobs. This desired economic result seldom occurs because too little attention has been paid to the actual skill levels, training requirements, or abilities of the local population. Because human resources are more important than natural resources or even location, initial planning must focus on current as well as projected labor force skills. An unskilled, or even technically but unsuitably trained, workforce can rarely meet the labor market conditions of new firms.

4. *Following the fad.* High technology, tourism, and convention centers dominate current municipal economic strategies. Communities adopt these trendy approaches on scant economic evidence and without any particular ability to provide the necessary infrastructure for these firms. A substantial body of recent research indicates that there is little support for most communities' aspirations for a high-technology, tourism, or convention scenario. In fact, unless a community has unique attributes, with the necessary reinforcing institutions (such as a university's relationship to high technology), prospects for such development are dismal no matter how many artificial stimulants or inducements are provided.

5. *Overlooking development capacity.* Too often, local governments or neighborhood groups have adopted development strategies and proceeded without assessing the overall development capacities described earlier. Groups should undertake serious examination of the capacity to develop and manage any project to avoid this pitfall. In many instances, planning efforts have failed. Development strategies and projects must be designed to fit not only the area's or region's resources but also its competencies. Indeed, the success of a local development strategy depends on a

long-term commitment by a sustained coalition of local public officials and corporate and labor leaders to carry out the program or activities.

Assembling the Elements of a Strategy

A strategy is a collection of actions and activities that help achieve a predetermined goal. The methods used to assemble the strategy, as depicted in Figure 6.1, are critical in building one that is long-term. The following factors form the base for assembling such a strategy.

Target Characteristics

Target characteristics include such items as the strategy's scope. Is it multi-jurisdictional, single-city, or neighborhood? The strategy may also focus on a specific sector for revitalization and emphasize certain types of enterprises such as small businesses or large export firms. In addition, the strategy can address the direction of economic growth by selecting new firms versus existing enterprises, or it may single out some elements of both. Finally, a strategy must address the critical environmental constraints that may impede its successful implementation.

Methods of Development

A good strategy uses appropriate methods to accomplish intended objectives. These methods might include direct assistance to firms, such as financial incentives, land assembly, and other tools. Or alternatively, the strategy may emphasize improving the procedures or processes available within the local environment for firms to establish or expand into new markets. The means to accomplish these objectives range from improving the existing permit procedures to business education and marketing assistance.

Local Development Institutional Form

In many instances, the form of the economic development organization is selected before a strategy is designed. As a result, the institutional form is not compatible with the strategy. For example, local governments are not very good vehicles for taking ownership of declining enterprises or for starting new businesses.

142

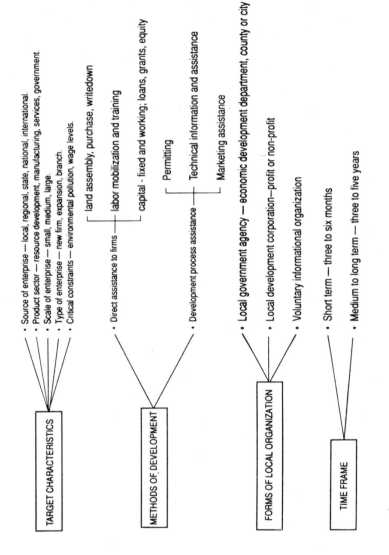

Figure 6.1. Elements of Strategies for Local Economic Development
SOURCE: Teitz and Blakely, 1985. Used by permission.

The organizational form needs to be well thought out before economic development strategies are selected. Voluntary organizations, nonprofit community development corporations, local development corporations, business associations, and neighborhood organizations all have strengths and many limitations when it comes to local economic development.

Time Frame

A local economic development strategy must include both short-term, visible objectives as well as long-term, process objectives. It is important that the local decision makers consider clearly how to incorporate both shorter-term as well as longer-term goals in any economic development strategy.

Projects from Strategies

Having developed a development strategy, the next step is to build an action plan for each of its viable projects. Action plans are documents that describe the components of a proposed project that match the economic development strategy. The principal purpose of the project action plan is to provide sufficient information to test the project's viability—that is, to determine whether the necessary economic, technical, management, and other support systems will indeed be adequate to support the proposed project. Taken as a whole, all the projects with detailed action plans constitute the community's overall development program.

The components of an action plan are shown in Figure 6.2. They include a statement of project inputs, a management structure and institutional plan, and a statement of project outputs. Each of these components is discussed below.

Describing Project Outcomes

Planning cannot commence without an examination of project outputs, because one cannot specify exactly either project inputs or a project management and institutional plan without knowing what the project is to produce.

The first task in building an action plan should be that of *specifying the goods and/or services* that the proposed project will produce and/or sell. All of the other factors in the action plan will relate to this definition. Other project outputs that should also be

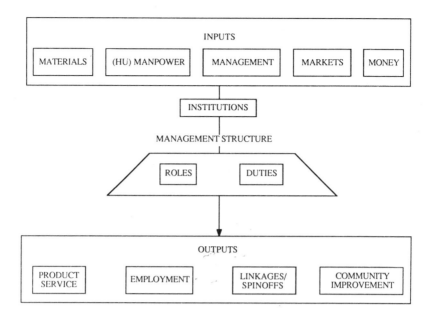

Figure 6.2. Components of an Action Plan

described at this time include employment impacts, linkages and spinoffs, and community improvements.

Employment Impacts

Job creation, in general, is important, but it is more important to create the right jobs for those in need in the community. As mentioned earlier, it is far easier to create jobs for the existing human resources than it is to move people to new jobs or to attempt to retrain them for other forms of employment. It is important, therefore, to ascertain the types of jobs created in relationship to the existing employment requirements of the community. Although this will necessarily be a rough estimate, it will provide a framework for ensuring a better fit of jobs to people.

A statement of employment impacts should include answers to at least the following questions:

- How many new jobs will be directly associated with the proposed project?

- In which occupational category and wage range will the new jobs fall?
- Will the new jobs provide better-paying jobs, better working conditions, or better career advancement possibilities compared to existing employment opportunities?
- How many of the new jobs are likely to be filled by local residents? How many by newcomers?
- How many of the new jobs for local residents are likely to be filled by unemployed people, women, or minorities? Would job training be required to accomplish this? If so, what kind?
- Will the new project result in the elimination of jobs, directly or indirectly, in other local businesses?

Linkages and Spinoffs

The extent to which new projects fit into the existing business and industrial base is an important consideration. If new businesses drive out existing firms, little is gained from the development planning process. However, some projects expand or improve the existing business base or even create new industry bases.

The attraction, expansion, or start-up of businesses or services that have linkages with existing economic activities in the region will have a more beneficial socioeconomic impact than industries with no local linkages. Socioeconomic linkages can include the following, for example:

- Use of local raw materials
- Use of locally produced goods or services as inputs
- Employment of local personnel (particularly if they have been locally trained)
- Sale of the goods/services produced to the local market
- Attraction of new investment to the area
- Attraction of ancillary employers to the area

Projects that strengthen socioeconomic linkages are usually also projects with a competitive advantage.

Community Improvement

Development planning is not merely (or solely) a business development activity. Culture, recreation, community appearance, and related factors are integral to development planning. As a result, each project should be measured for its contributions in these areas in addition to the areas of direct employment and economic development. For example, projects that build the self-esteem and identity of youth can significantly alter a community's social and economic climate. These aspects of a project, which enhance the total livability of an area, should be documented under the heading *community improvement.* To reiterate, the specification of outputs for proposed projects forms the definition and guide for the entire development process. One cannot determine other project requirements without discussing with knowledgeable people the full potential for success and failure. Similarly, platitudes about job creation or improving the overall economy are meaningless. The real focus for economic development comes from trying to work through specific activities that improve actual situations using known or attainable resources.

Specifying Strategic Resources: The Five M's

Economic growth and development can be seen as analogous to a living plant's growth and development: The key elements of heat, light, water, and other nutrients must be present in proper amounts for a plant to grow. If one key element is missing, growth will not occur, even if other elements are present in abundance. Similarly, five key elements, termed the *Five M's,* must be present for economic development to occur. These elements are (1) materials, (2) (hu)manpower, (3) markets, (4) management, and (5) money. Each must be considered by communities in creating an economic development plan. This section defines these elements (see Table 6.1) and discusses their importance to economic development programs.

Materials

Materials may be thought of as all existing physical resources, including both natural resources and human-made facilities or infrastructures such as roads, ports, electric power distribution systems, and buildings. Natural resources, of course, encompass all useful, naturally existing materials and conditions in such forms as soil, terrain, minerals, climate, water sources, plant and animal life, and geographic locations. A community should understand that although these resources may be limited in quantity and availability, they may be combined in any number of ways to produce a wide variety of products and services. Because absence of materials can restrict the types of goods that may be produced

Table 6.1   The Five Key Resources for Economic Development

Materials
- Land
- Buildings
- Location
- Infrastructure/natural resources

(Hu)manpower/Labor
- Skilled personnel
- Available workforce
- Education and training capacity

Markets
- Markets analysis
  - Competition
  - Penetration
  - Marketing strategy

Management
- Organizational structure
- Managers/operators
- Research and development (R&D)
- Marketing and sales
- Legal

Money
- Equity/ownership capital
- Debt/borrowed funds
- Capitalizing institutions
- Subsidy and substitutes for direct capital

in an area, however, the existence of these resources should be investigated before attention is given to potential products and their production in the effort to provide jobs.

(Hu)manpower

(Hu)manpower represents the labor used to create a product or service for sale. For example, labor includes the person operating a machine, repairing a machine, supervising the process, or delivering the final goods produced. Specific skills may be needed for different stages of a process, and those skills may be acquired through experience or training. These skills amount to an investment that has a clear "payoff" in the form of improved productivity. A trained and experienced person will be more disciplined, make fewer mistakes, and be more innovative. The skilled person is, therefore, an asset and can be thought of as human capital within the economic development process.

The amount of human capital available for work in an area depends on the size of the working population and people's willingness and ability to work. Willingness to work depends on the sociocultural background of the individuals, the type of work available, and the prevailing wage rates. Ability to work depends on the age of the population and the level of education and training. Existing (hu)manpower in an area is a significant resource.

Markets

Markets are the places where there is demand for certain products or services. Demand size depends on the number of people or organizations desiring the product; the qualities of the product; the price at which it is offered; and the ability to inform possible consumers of the product's quality, price, and availability. The market area for a product is not fixed. A change in population characteristics, the local or international economy, or simply a change in general social values and attitudes may create potential customers. Similarly, new methods of communicating price and product characteristics create the potential for new demands. New markets may be created through better product pricing, superior transportation, better production techniques, and/or penetration of more local or overseas markets. The point is that markets are normally very flexible and can be expanded through imagination and hard work. Identifying potential consumers, knowing what kind of product they want and at what price, and being able to inform them of the product's availability are examples of using a market area to maximum effect.

Management

Management is a special type of human capital. A good manager is the person able to combine materials, money, and personnel in order to produce and market a product successfully. Effective management is the catalyst without which economic development cannot occur. The most effective managers also tend to be visionaries, risk-takers, and innovators, as well as motivators and coordinators.

Economic development is a multifaceted process and, for it to be successful, there must be effective coordination between government and industry as well as between various agencies and levels of government. Most important, the local community must provide support and involvement. Successful links among these diverse groups require top management talent in both the public and the private sectors of the economy. Local government or other initiators of economic development programs designed to stimulate employment may have to be especially creative in finding ways to contribute to the availability and use of effective management talent. The management needs of ventures are discussed further in a later section of this chapter.

Money

Money is the financing directly involved in establishing and operating a proposed project. Money is required to initiate an economic development project in order to provide necessary facilities, to hire and train staff, to pay for materials if a product is to be manufactured, to transport the product, and to market the product or service.

When there is a need to provide missing elements or components for a successful project, money must be found to purchase, hire, organize, or in some other way make these elements available. For example, funds may have to be found to acquire the use of buildings, hire accountants or management consultants, train workers in specific skills, or extend sewer lines and roads. Funds for the necessary project components may come from private or public sources, and the components themselves may be provided by public or private agencies. In the next section, I will describe a variety of alternative approaches to finance.

Project Finance

Money, the last of the Five M's, causes the greatest problem for most local governments or development organizations because it is too often thought of as "funding" rather than "financing." In the public sector, funding refers to obtaining all of the fiscal resources to develop and manage a project or program. In essence, government agencies usually think of a continuing commitment of tax dollars to support a public activity regardless of any income generated from it. However, *financing* refers to identifying sources of capital (usually private) to provide the initial financial resources. It is further anticipated, based on careful analysis, that the project will generate sufficient income to pay for itself.

Local development organizations need to think in terms of financing projects rather than funding them. Project financing requires the flexible use of existing assets combined with solid project planning. For example, city land can usually form the basic fiscal resource for financing a local agency's building project. Similarly, local governments have become adept at selling development rights for the construction of needed office space above parking structures or other community facilities. Some cities have used civic assets such as their museums, city halls, and other facilities as their contribution to commercial and retail developments in which the city government is a major investor. Many cities have used a lease agreement for the major portion of a commercial structure, thus encouraging a private developer to build a new multistory administrative building and civic library. In other words, project financing requires imagination more than actual money. A well-documented economic demand for a service or the use of city government space, development rights, or other factors can assist the community in achieving its objectives without using local taxes. Project identification follows the assembly of the appropriate local fiscal resources to meet the community's goals. Good project ideas are the key to development.

In a typical situation, identifying economic development projects will require compromises and trade-offs. There may be projects that improve people's well-being, increase productivity and economic viability, or enhance the quality of life and work—but there may be no markets available for such products. An economist would say that wants and needs are greater than effective demand (that is, the ability to pay). In the typical situation, there would have to be trade-offs between meeting local development needs and making money. Thus, the most general guideline for identifying projects is to look for those that enhance local economic development and generate acceptable profits, simultaneously, at reasonable levels of risk. An obvious but important implication of this guideline is that local communities should avoid at all costs projects with high risk and/or no profit. If an acceptable cash flow cannot be anticipated from a potential project, the project should be discarded. With the criteria of economic development potential, profitability, and risk in mind, these guidelines form the minimum mean test for a community's participation.

Summary and Conclusion

A set of conceptual parameters for local economic development planning has been laid out here, forming its basis—strategy development and program design. The planning system or orientation taken by any community will shape the goals of the process and the resources used.

The collective experience of many hundreds of local areas working to develop their economies will shape new strategic models. Unavoidably, perspectives favoring corporate and local needs will exist side by side in the same locality. Similarly, nearly every area comprises some growing industrial sectors and others that are declining. Thus, these areas will exhibit simultaneously two or three requirements for economic development. Communities cannot remain as complacent or indifferent to questions of economic development as they did a decade ago. The careful economic development planner will sort through the economic development models outlined. Strategies emerge from the circumstances. None will conform to ideal types depicted here. In the following chapters, these strategies are explored in greater detail and combined with case studies that illustrate how they are integrated into a community environment.

References and Suggested Reading

Barnes, Nathan et al. 1976. *Strategies for an Effective Public-Private Relationship in City Industrial Development.* Prepared for the Society of Industrial Realtors. Washington, DC: Economic Development Administration (NTIS).

Bendavid-Val, Avrom. 1980. *Local Economic Development Planning: From Goals to Projects.* Report No. 353. Chicago: American Planning Association.

Bluestone, Barry, and Bennett Harrison. 1980. *Capital and Communities: The Causes and Consequences of Private Disinvestment.* Washington, DC: Progressive Alliance.

Carlisle, R. 1978. New Strategies for Local Economic Development. *Carolina Planning* 2 (4): 14-18.

Gardner, L. M. 1983. *Community Economic Development Strategies: Creating Successful Businesses: Vol. 1. Building the Base.* Berkeley, CA: National Economic Development and Law Center.

Kelly, R. 1976. *Community Participation in Directing Economic Development.* Cambridge, MA: Center for Continuing Economic Development.

Kemp, Roger A. 1992. *Strategic Planning in Local Government*. Chicago: American Planning Association.

Mahood, S., and A. Ghosh, eds. 1979. *Handbook for Community Economic Development*. Sponsored by the U.S. Department of Commerce, Economic Development Administration, Washington, DC. Los Angeles: Community Research Group of the East Los Angeles Community Union.

Malizia, E. 1981a. *Contingency Planning: A New Approach to Local Economic Development Planning*. Unpublished paper.

Malizia, E. 1981b. *A Guide to Planning Economic Development in Small Communities and Rural Areas*. Charlotte, NC: Division of Community Assistance, North Carolina Department of Natural Resources and Community Development.

Malizia, E. 1981c. *Planning Economic Development in Smaller Communities*. Paper prepared for the National Planning Conference, Rural and Small Town Planning Division Session, Boston, April 28.

Nathanson, J. 1980. *Early Warning Information Systems for Business Relocation*. Washington, DC: Urban Consortium, U.S. Department of Commerce.

National Council for Urban Economic Development (NCUED). 1977. *Strengthening the Economic Development Capacities of Urban Governments*. Washington, DC: Author.

Schmenner, R. 1980. How Corporations Select Communities for New Manufacturing Plants. In *Firm Size Market Structure and Social Performance*, edited by J. Siegfried. Washington, DC: Government Printing Office.

Teitz, M., and E. J. Blakely. 1985. Unpublished course materials. University of California, Berkeley.

Tremoulet, A., and E. Walker. 1980. *Predicting Corporate Failure and Plant Closings: Resources for Local Employment Planners*. Unpublished working paper. Chapel Hill, NC: University of North Carolina, Department of City and Regional Planning.

7 | Locality Development

The term *public/private partnership* has entered the permanent lexicon of local government. What is a public/private partnership? Why is it important to consider it as a component of economic development?

Public/private partnerships are not a new phenomenon. They are the legacy of over 50 years of federal urban policies. In 1938, the federal government embarked on a set of housing assistance programs by chartering the Federal National Mortgage Association (FNMA), now commonly known as Fannie Mae, to create a secondary market for home mortgages. This program created a partnership between the public sector and the private market to produce housing in urban areas. The cities benefitted from this national partnership. It was so effective that in the 1960s and 1970s this concept was extended to rebuilding the inner cities through the Model Cities Program and later the Urban Development Action Grant (UDAG). UDAG was the most powerful stimulus for downtown restoration ever devised. Most of the nation's large cities cleared large portions of their downtowns to open them to revitalization projects crafted at the local level but supported through federal government transfers and loan guarantees. The UDAG program established the framework for local partnerships between government and local private sectors.

In the mid-1970s, local officials began experimenting with a new set of relationships with the private sector in order to complete the projects launched a decade or more earlier under more generous federal support. The lessons learned from UDAGs were adapted by cities with their own funds in the 1980s when federal support was withdrawn at the height of local fiscal distress.

City officials became dealmakers by the late 1970s. These deals included the provision of essential infrastructure at little or no cost with promise of a return on the city's investment through a soft loan or a portion of the profits from the project. In many instances cities built public garages for private developers or sold or leased back facilities to retailers. In essence, cities moved from their traditional role as regulator to coinvestor. "This new role marked a change," as Sagalyn says:

> in expectations about the public sector's separateness from the public sector. . . . These strategies placed a high premium on public entrepreneurship and private market feasibility. . . . And increasingly, in bargaining with developers, cities negotiated direct financial stakes in a project in the form of public profit sharing arrangements, a practice very unlike the usual behavior of cities. (Sagalyn 1990, 429-430)

As a result, most local governments entered the 1990s with considerable experience as entrepreneurs in the development process. Land management and land deals are forming an important component of any economic development program. Land is one of the most important factors in local economic development today. Without control of land, local development is essentially impossible. A local or community development plan will be thwarted if suitable sites and/or buildings for selected projects cannot be furnished. Land must be carefully managed. This management must aim to use land prudently and improve existing land-use practices and the appearance of the community. One often overlooked key factor is the design standards associated with urban space.

As stated earlier, livability is an extremely important component of economic development. An attractive entrance and exit for a city tells the visitor a great deal about the pride and community spirit of its residents. Moreover, both residents and nonresidents tend to spend their time and money in the most pleasant of the available environments. Therefore, tree-lined streets, covered walkways, and clean, well-managed shops can help build local business, either by retaining local spending or by attracting customers from outside the region. Indeed, there are few factors in tourism more important than community appearance. Visitors do "judge the book by its cover." In fact, judging cities by their appearance and their social/economic climate has become a major assessment tool of economic development professionals. *Fortune, Money, Finan-*

cial World, and other leading business periodicals present annual assessments of civic performance on the basis of the local capacity to do business with the private sector along with the physical and social appeal of the community to its indigenous population and visitors. One result of these polls is the transformation of the desirability of communities such as Pittsburgh, Seattle, Sioux Falls, Baltimore, and Oakland, California, over the last decade. The Rand McNally quality-of-life survey is viewed by some economic development practitioners as a legitimate locational preference index for corporate real estate managers.

Local officials find it increasingly important to do "image management" in the face of global competition for corporate location. As a result, they must carefully develop their land, building assets and "presentation features." The motivation behind developing land management and control systems should not be to prevent the worst things from happening but rather to get the right things to happen. This requires land-use planning that goes beyond coloring maps and issuing regulations. It calls for the development of a visual theme that (1) creates a sense of identity, (2) improves the amenity base or livability of the community, and (3) improves the attractiveness of the civic center in an effort to improve local business. Table 7.1 reviews how the most commonly used locality development tools achieve these ends. Each of these tools is described below.

Landbanking and Community Land Trusts

Landbanking is the practice of acquiring and improving contiguous parcels of land. Local development organizations can use land banking to put together good development sites for business or industry. Landbanking does not necessarily mean vacant site acquisition. In fact, some of the most creative landbanking is accomplished by local governments. A number of cities have rented or leased land or buildings acquired through tax sales or eminent domain to assist in starting up new businesses or developing new markets. For example, some cities have leased land acquired through street widening or located under freeways to new environmental enterprises. In other instances, cities have reused vacant warehouses as community ice skating or rollerskating rinks prior to razing the facility for another use. In each of these cases, the community has used an underutilized asset to establish a new

Table 7.1 Location Development Tools and Criteria

| | Objective | | |
Tools	Image Building	Amenity Improvement	Business Improvement
Landbanking		X	X
Infrastructure provision		X	X
Speculative buildings			X
Incentive zoning		X	X
Regulation improvement	X		X
Tourism planning	X	X	X
Townscaping	X	X	X
Shopsteading	X		X
Housing and neighborhood improvement	X	X	
Community services	X	X	X

enterprise that can be relocated when the land or buildings are converted to new uses. Nothing is more unsightly than vacant and derelict buildings.

Landbanking is a powerful locational incentive, done primarily in older communities where little land is available for development and also in newer communities to reserve good sites for future development. Landbank sites often include land that is surplus city-owned, donated, acquired through condemnation, purchased from private sources, former stockyards, or former military installations. To build a landbank, a locality should set up a real estate division to search continuously for underutilized and/or underdeveloped properties, catalogue these properties by size and location, and computerize the information for rapid updating and quick reference.

At the neighborhood or community level, a community land trust offers a unique means for communities to maintain control over their physical assets. Community land trusts are local groups that establish themselves as nonprofit land trusts. The trust board and leadership is generally composed of local neighborhood activists, businesspersons, and relevant professionals (planners, architects, finance and real estate brokers). The initiating group can call on technical assistance from the Institute for Community Economics in Springfield, Massachusetts.

The most common community trust usually acquires vacant homes through a loan, gift, or city reconveyance in blighted neighborhoods or in areas undergoing significant disinvestment. The trust identifies low or moderate income families and assists them in rehabilitating the property. The families' "sweat equity" may be used as down payment for the house. The trust sells the home but maintains a land lease on the ground that requires owner occupancy and limits the resell price on the property. In some instances, state housing financing agencies offer the mortgage loans, and in others, banks or bank intermediaries fulfill their Community Reinvestment Act quota through such loans.

Community land trusts have expanded to include the purchase of farmland from hard-pressed farmers, re-leasing the land to the farmer under the condition that he/she maintain the land in agriculture.

Landbanks or community trusts require substantial capital. Potential sources of funds for land acquisition and preparation include national, state, or local capital programs, Community Reinvestment Act funds, program-related investments of foundations and intermediary institutions such as the Local Initiative Support Corporation of the Ford Foundation, and industrial development bonds. If money for land acquisition cannot be found, however, an alternative approach is to gain control of suitable parcels of land by purchasing an option on them and later selling the option and land to a suitable developer.

Physical Infrastructure Development on Industrial and Commercial Land

Industrial and commercial land and buildings are often more attractive to potential businesses and industries if they have already been improved. The advantages to a company buying or leasing improved land are twofold: the time between acquisition and operation can be greatly reduced, and the expense and bother of site improvement are avoided. Thus, many towns and industrial development corporations have the requisite building inventory and physical infrastructure already in place to attract new companies and retain existing, expanding companies. Some have been successful, others have not. When considering this tool, be sure it is only used in relation to businesses truly suited to an area.

Adaptive reuse is another alternative for community infrastructure revitalization. A number of cities have developed policies to reclaim old underused industrial property for other uses. Factory space has been transformed into housing, live-work space, offices, art studios, restaurants, and shopping malls. The city usually uses its resources to reconfigure the streets, build parking structures, and reinforce the building to meet basic requirements of the new users. Federal and state rehabilitation tax credits allow the developers to receive tax breaks for a period of 5 or more years for the conversion investments.

There are a variety of additional ways in which a community can improve land and/or buildings to create incentives. Most common is the provision of water and sewer lines, street lighting, access roads, and sidewalks. Lowell, Massachusetts, converted 11 major textile mills into office and demonstration space for a national urban park, with the city providing off-site improvements and parking. Others cities following the Lowell example, such as Charlestown, South Carolina, have provided landscaping for privately financed historic preservation efforts of industrial, commercial, and housing estates. The Tennessee Valley Authority provided technical assistance to small rural communities for the transformation of old city motels into residential space called "hometels" (Sagalyn 1992, 7).

Speculative Buildings

Speculative buildings are "shell" buildings. Their interiors are left largely unfinished until a tenant is found. They are a marketing tool for attracting firms to an area and/or for retaining existing companies that are expanding. They provide work space—a key factor in a firm's decision making on site location. By providing the space, a city or neighborhood can significantly reduce a firm's start-up or expansion time.

Speculative buildings are best used by localities that suffer from a shortage of industrial space but that have an adequate labor force, a transportation system, and a sufficient supply of public services including utilities, police, and fire services.

One inventive and low-risk approach to speculative building is for a locality to build fully serviced storage units with water and power. These units, if properly designed, can be turned into small factoryette space, repair shops, and business incubators. A number of communities in the Northwest have either assisted in such space development or converted their own underutilized corporate yards to this type of incubator space. Perhaps the best example of this type of investment in the United States is the very successful biotechnology incubator space provided in Emeryville, California, adjacent to the University of California at Berkeley.

Zoning Regulations

Zoning policy can promote economic and commercial development by setting aside a sufficient amount of land for industrial and commercial use and by allowing flexible zones and rules in the local zoning code. A number of techniques have been developed that overcome deficiencies of single-use, strict limit designations of height and density—thus making land use more flexible. Such zoning tools include incentive zoning, overlay zoning, and special districts. Because these regulations are controlled by the local government, only local planning and zoning officials can use these tools.

Incentive zoning is often used to overcome strict site regulations of height and/or bulk. It provides a developer with flexibility and encourages certain land uses and project features. These incentives may be applied in a variety of ways, but they are generally used to obtain public benefits in exchange for design concessions to a developer. The most prevalent type of incentive zoning is bonus zoning, in which additional densities or increased floor areas, beyond those specified in the zoning codes, are awarded in exchange for public benefits (access to mass transit facilities, open space, and so on). This system of zoning generates its own set of problems because there are no fixed standards. The City of Chicago established a manufacturing district in 1988 in an effort to preserve production-related activities in the city limits. The incentives in this zone included a restriction on housing, schools, and community facilities in the district as well as other incentives designed to appeal to manufacturers.

Overlay (or floating) zoning relaxes static traditional zoning that assigns specific uses to particular land tracts. Instead, overlay zones are a special modification to the existing basic land use provisions. An overlay zone may contain regulatory provisions that designate land uses, height, and bulk as a standard zoning ordinance, or it may have unique features such as historic preservation incorporated into the ordinance that provide for a specific purpose, such as an industrial park or a mixed-use development. In addition, overlay zones can be used as a device to market development rights by selling to developers the right to increase plot ratios in a specific area while maintaining the same overall citywide density limits.

An excellent example of the use of overlay zoning is the historic corridor district developed in Gastonia, North Carolina. Gastonia's overlay zone is a supplemental zoning area bordering its historic residential district. The overlay zone protects the underlying historic district residential character and design, and the overlay provides a development corridor in the area that allows professional offices, banks, specialty shops, and small restaurants in keeping with the overall historic theme. As a result, commercial development and residential character were both enhanced.

In essence, overlay zoning merely allows the shifting of densities and zoning applications around the city within prescribed limits. It also allows the market to work more effectively and efficiently. Developers can place facilities where the demand exists rather than where planners think it ought to be. Finally, in growing areas, this tool is a tremendous fund-raising device, allowing the local government to acquire capital for other civic improvements without going back to the taxpayers. At times, this practice alone generates sufficient income to finance the community's economic development office and/or program.

Regulatory Improvement Through Simplification

An excellent method of evaluating a community's regulatory system is to place a local public official in the developer's role and have him or her walk through a typical permit or code enforcement routine. This will provide the local official with insight, from a developer's point of view, into any flaws and problems associated with the development approval process, such as conflicting regulations, time-consuming delays, or a negative or adversarial view of development. If this procedure indicates procedural and regulatory shortcomings, the community should alter its regulations and simplify the approval process. The Loma Prieta earthquake in San Francisco in 1989, the Oakland, California, firestorm in 1991, and Hurricane Andrew in 1992 all provided local government officials with intense and expensive reviews of their existing zoning procedures. The aftermath of each of these events provided local officials with a better understanding of how their planning and building systems were ill-equipped to cope with the development process.

Recently, many communities have established local interdepart-
mental panels, one-stop permit offices, or committees to review
development impacts and overcome the aggravation and uncertainty involved in the process.

Tourism Planning

The development of a detailed physical improvement plan for
tourism is generally an afterthought by local officials. Frequently, local
residents complain about tourists parking on residential streets or
coming to their homes and businesses merely to ask directions.
Careful planning can avoid these simple problems.

One technique is to mark off *sacred community structures* and
make them inaccessible to the tourist. For example, the local
swimming hole or a particular stretch of beach may be perceived
by the community as a local venue, inappropriate for tourist use.
Access to these areas can be limited by simply not putting up signs
to direct outsiders to them. Alternatively, clearly marking tourist
sites and developing well-marked roadways to direct tourists to
designated sites are extremely helpful. Some communities have
marked the roadways with specially colored signs so that the
tourist can quickly identify the correct routes and not cause congestion on local streets. Johnson County, Tennessee, the home
territory of Daniel Boone and "moonshine whiskey," failed to
market itself to tourists until 1986, when the local hospital went
bankrupt. At that point the residents decided that they needed to
form a tourism committee and market their community. The first
order of business was to "dress up for company" (Walsh 1992, 1).
After dressing up, the community developed a realistic tourism
agenda that resulted in a new community visitor center, the subsequent reopening of the hospital, and a textile plant. Tourism brought
industry to Johnson City.

Townscaping

Many small towns believe tourism is the antidote to collapse of
the town center. Although tourist dollars are important, local trade
is important as well. Tourism is both cyclical and fickle. Commercial centers must ultimately survive on local traffic. Communities

must therefore do everything they can to make shopping in their civic center attractive and worthwhile for local residents as well as for outsiders. One of the best ways to make the downtown area attractive is to embark on programs that give it character or restore some character to it.

Townscaping is one means of achieving these objectives. Townscaping is a physical, attitudinal, and management process. The physical component basically entails the development of a visual theme for the central town area by local merchants, city planners, and citizen groups. The theme may be based on the area's history or be a more contemporary expression of the community's identity. The *Main Street* program is the usual manifestation of this effort. In this program the town theme is incorporated into plans for building or rebuilding the existing landscape and built environment. Some small communities in the Midwest have developed modest but well-designed approaches to townscaping that make them exceptionally visually appealing. Many small communities have experienced increased local business traffic and tourism as a result of improving their visual image while preserving their rural or small town character.

The attitudinal component of townscaping relates to the actions local businesses take with regard to their town center. Some communities have resurrected or formed new local chambers of commerce to provide business with a forum in which to discuss common problems such as improving the skill level of shop assistants and increasing the type and range of advertising on television or radio as a group advertising effort.

The management aspects of townscaping include such measures as hiring consultants or even civic center managers to help analyze methods for increasing sales, improving product offerings, maintaining cleanliness and attractiveness, and advocating civic improvements that will improve retail opportunities.

Shopsteading

Shopsteading is a relatively new approach to community or inner-city neighborhood revitalization. It is a tool that can be used to address the problem of vacant commercial property. It involves the sale of abandoned shop facilities to businesspeople willing to renovate them and then operate their businesses there. The success of shopsteading hinges on two main factors, the first of which

is the availability of vacant properties in areas that have considerable potential for economic rejuvenation. This means that there
must be an identifiable market for goods and services in the
shopstead area. Typically, these are areas undergoing "gentrification" (i.e., new, young families or single people moving into the
area and rehabilitating old dwellings). The second necessary component is the availability of individuals able to satisfy shopsteading
requirements. Before settling on the property, shopsteaders are
usually required to provide evidence of equity capital and to submit
detailed specifications for rehabilitating the property. They also
must have estimated all capital requirements for reconstruction
and obtained commitments for the necessary financing. Shortly
after settlement, the shopsteaders must begin improving the
property. Within a few months, the building must comply with
the local building code; within a year, the shopsteaders must
have completed the renovations. If not, the property reverts back
to the city. Within a year, the shopsteaders must begin operating
the business and continue in operation for at least 2 years before
it may be sold.

Shopsteading is designed to promote both business retention and
business attraction activities. For businesspeople, the most important
benefit is low-cost, sometimes rent-free property; the shopsteaders
have the immediate advantage of purchasing property at a cost
considerably below the market value. Owning the building where
his or her business is located gives the small businessperson an
incentive to improve it. Shopsteading probably provides one of the
best opportunities for people who wish to open retail and specialty
shops, service-oriented firms, and other small businesses.

For the city or town, shopsteading has the potential for improving land-use patterns in marginal commercial areas and providing
incentives for other private investors to renovate the remaining
buildings on the block. Before deciding to initiate this program,
however, local governments and community groups must study the
opportunity costs of shopsteading (how the available resources
might be used for another project).

Housing and Neighborhood Improvement

Housing is one of the most obvious community resources, yet it is
rarely considered by most cities with regard to economic development. There are two aspects of the housing-economic development

nexus. One is the need to provide diverse housing types to provide homes for all groups in the community. The other is the need to provide households with services ranging from child care to community facilities such as swimming pools. The traditional approach taken to these opportunity areas has been merely to wait for the state or someone else to act. City councils, however, are becoming increasingly aware of how the availability of housing and household services influences the local economic development climate. As a result, some cities are becoming aggressive housing developers through the use of their land. BRIDGE Housing Corporation in San Francisco is the largest nonprofit housing producer in the nation. BRIDGE produces over 1,000 units of housing annually both for sale and rental. BRIDGE developments are based almost entirely on local land contributions to support the project. There are over 1,000 nonprofit housing development corporations in the country that pursue similar strategies.

Housing development is not an expensive process if a community is prudent about land acquisition. By purchasing or trading land, community groups or city governments can put themselves in the position of inviting quality developers to build diverse housing. Land cost is a critical determinant in whether or not good affordable housing can be provided.

Household services represent another area in which there are new opportunities for enterprising local governments. Some councils have built new recreation and sport facilities under lease agreements with private operators rather than providing and maintaining these facilities themselves. Community-based organizations can provide household and community services on a contract basis for neighborhood development and job generation.

Community Services

Local government provides a wide range of services, from household to commercial. Communities are also becoming involved in providing services such as running trailer parks, operating summer youth projects, writing local histories, and maintaining museums and art galleries. In other instances, councils are beginning to operate motion picture theaters and similar labor-intensive services. These services are frequently a drain on city resources. There are ways, however, for communities to reduce these costs and, in some instances, turn them into money-making propositions.

Local government and community-based organizations have used a number of techniques to make community services pay for themselves. Local governments have sublet their visitor centers to entrepreneurs who run them as combination gift shops and information centers. In addition, as communities embark on tourism as an economic activity, they discover that tourist services must pay for themselves directly. Visitor bureaus and similar activities are expensive. There are ways to finance these services, however. For example, some visitor centers publish tourism newsletters with advertising for local merchants. This helps defray some of the cost of the center.

The approaches communities use to provide services while reducing costs and creating jobs is limited only by imagination. The traditional fiscal municipal management philosophy of "we won't do it unless we've got the money" is no longer acceptable. For example, communities in Florida and Maine aim at attracting tourists willing to pay more to "sustain" the natural environment. Local officials have the responsibility of finding a way to do it rather than a reason for not doing it. Recognizing this, a local government should develop alternative mixes of public and private resources, combined with incentives, to achieve its objectives. Similarly, community groups cannot take "no" for an answer. They must also develop a strong blend of existing and new resources to meet their goals. The following projects and case studies represent means by which a community can utilize its civic assets.

Illustration 7.1. Riverfront Gamblers

Cities of all sizes have learned to ignore their waterfronts at their peril. Not only do these areas become eyesores, they discourage business. But things have changed in the last decade—and not only in big cities. Small communities in North Carolina, Minnesota, Colorado, and Washington State, among others, are leveraging private investment to spruce up their waterfronts and bring vitality back to their downtowns.

Leveraging Redevelopment

By 1979, suburban malls had siphoned off the major retail stores in Wilmington, North Carolina (pop. 57,000), while downtown increasingly was left with the empty hulks of old, abandoned buildings. The city decided to fight back by commissioning a study of ways to revitalize its downtown and adjacent riverfront.

As it turned out, the city had substantial untapped resources, according to James Hayden, principal in charge of the Wilmington office of the Fort Lauderdale consulting firm of Edward D. Stone, Jr., and Associates. The city's harbor on the Cape Fear River, the source of one-fourth of the area's jobs, dates back to colonial times and served as a key port for Confederate blockade runners during the Civil War. A sizeable historic district offered attractive residential and commercial buildings with renovation potential. And nonprofit groups were prepared to take the lead in promoting reinvestment strategies.

The centerpiece of EDSA's plan was a pair of waterfront parks at the northern and southern ends of the central business district to create a "more pleasing environment for economic activity," according to Michael L. Hargett, the city's riverfront project director. The central idea, Hargett notes, was to invest in public facilities to "leverage private investment downtown." In 1982, the city's planning department followed up the consultant's recommendations with a plan that identified 20 specific facilities for redevelopment.

The city was not starting from scratch. In the late 1970s, developers had rehabbed the old Cotton Exchange warehouse, using its historic theme to create a specialty retail shopping complex and breaking with such urban renewal trends as slick facades. Its success, says Hargett, made it the riverfront's northern anchor and generated momentum for further commercial investment in historic buildings.

In 1982, the city marshalled over $1.15 million, almost entirely in federal funds (a combination of Urban Development Action Grant, Community Development Block Grant, and Heritage Conservation and Recreation Service monies), to launch work on the riverfront parks. In reality, the south park is as yet only a boat ramp with some open space, Hayden notes, but the city built a single major riverfront park combining space from two of the recommended facilities by adding $1 million for extensions and a riverwalk. The new parks, with a turn-of-the-century design theme, have enhanced the value of private investments, Hargett notes, and coupled with renewed economic growth they have kept the entire riverfront program on track. All of the 1979 EDSA schemes have been completed.

With the declining availability of federal funds, the city has sought new financing for its projects, turning to cooperative agreements with developers for sections of the riverwalk. In May, the new owners of a Hilton hotel on brick-paved Water Street decided to stabilize the riverbank behind the hotel parking lot. The city forged an agreement that would extend the riverwalk with a 920-foot-elevated boardwalk across the property. To fund the project, the city applied to the state for a $100,000 shore-front access grant and arranged to repay the Hilton another $500,000 through a five-year installment purchase agreement. The city is to take possession of the boardwalk upon completion of the improvements.

To date the city has also made 24 low-interest loans from a revolving fund of roughly $275,000 to encourage facade renovations, according to city manager William Farris. Strict zoning prohibits nonresidential uses in the residential part of the historic district and requires certificates of appropriateness for exterior changes in commercial buildings. Last year, the city used an $11 million UDAG to help finance the mixed-use renovation of two former railroad warehouses and the construction of a 51-room inn by Coast Line Associates, a private partnership.

The $6.5 million complex includes a convention center and retail and office space and houses the Downtown Area Redevelopment Effort, a nonprofit group that has spurred several such projects. The city also gained a southern anchor to its riverwalk with the private restoration of Chandlers Wharf, a former warehouse turned into a tourist-oriented specialty shopping complex with a nautical theme.

Wilmington draws an estimated 1.2 million tourists yearly, 47 percent of whom visit downtown, according to Robert Murphrey, DARE's executive director. Among the draws: the permanently docked U.S.S. North Carolina, across the river from downtown but just a DARE ferry-ride away.

Murphrey notes that Wilmington's downtown retail sales nearly doubled between 1980 and 1987, from $27.5 million to $50.3 million, and that downtown investment totaled some $100 million in the decade ending in 1987. And DARE itself recently proposed Riverfront Center, which would add four levels of housing and street-level shops to an existing 500-car parking garage. DARE has already received some expressions of interest from potential major tenants, Murphrey says.

SOURCE: Schwab 1989. Reprinted with permission from *Planning*, copyright © Sept. 1989 by the American Planning Association, 1313 E. 60th St., Chicago, IL 60637.

Illustration 7.2. Unconquered Spirits

Flickering to life as the autumn dusk descends, a neon spire atop the Guadalupe Theater bathes one of San Antonio's oldest and poorest neighborhoods in green light and aquamarine. Although the glow begins to fade within a few blocks, the art deco playhouse serves as a beacon to artists and art lovers throughout the Americas.

It is a November evening 500 years after Columbus first reached these shores, and the theater is filled to capacity for the world premiere of *Cronica*, by Colombian playwright Enrique Buenaventura. Set on the Yucatan coast in the early 1500s, the drama explores the life of a shipwrecked Spaniard, his assimilation into Mayan society, and his ambivalence about being "rescued" by the Spanish explorer Hernán Cortés. Many in the audience—Chicanos who trace their ancestry to each of these cultures, and live within a third—respond knowingly to the questions Buenaventura raises about identity.

"Growing up in the barrio, we were never taught to see ourselves as part of a culture," says Enedina Vasquez, a local artist whose works have recently been exhibited in the Diego Rivera Museum in Mexico City and the governor's mansion in Austin. "We never saw that we had our own art, our own theater, our own music. Now we have a playhouse on our street corner again. We have studios to work in. We have instructors to teach our children. I'm telling you, this is a renaissance."

Credit for this cultural reawakening is due largely to the Guadalupe Cultural Arts Center, a local arts organization that attracts 150,000 visitors each year. Many of them come to see the center's resident theater and dance companies perform at the historic Guadalupe Theater, while others are drawn to its annual cinema, music, and literature festivals. Still more take part in some of the hundreds of educational workshops offered each year in such disciplines as poetry, printmaking, and percussion.

Like good theater, the center means many things to many people. To residents of the barrio, it is a catalyst for economic revitalization. To the San Antonio arts community, it is a generator of revenue and respect. And to Latinos throughout the United States, the center is a haven for Mexican-American culture, a training ground for emerging talent, and a leader among community arts organizations.

With more than half a million Mexican-American residents, San Antonio has one of the largest Latino populations in the United States. A few are descended from the city's original stock—Mexicans who lived there before Texas became independent from Mexico in 1836, or was annexed by the United States in 1848.

El Barrio Guadalupe, where the greatest concentration of Chicanos can be found, lies on the west side of town, two long miles from the downtown hotels, shopping malls, and esplanades that sustain a $2 billion-a-year tourism industry. The nucleus of the city's Mexican-American community for generations, the barrio is a neighborhood steeped in history, fraught with poverty, and imbued with hope.

Local legend has it that the barrio was established more than a century ago by cattle drovers who prodded their herds down Guadalupe Avenue toward a nearby sprawl of stockyards, slaughter houses, and meat-packing plants. Commercial development picked up after the Mexican revolution in 1910, when middle-class professionals and entrepreneurs fled to the United States and settled in San Antonio. With the onset of the Great Depression, however, many of these businesses failed.

The barrio has experienced its ups and downs ever since, and from the time the Guadalupe Theater opened in 1940 there has been no better barometer of the area's economic and cultural climate. During its first decade, when local residents found abundant work at nearby military bases, the theater was a bustling movie house showing films by such directors as Luis Buñuel. Attendance began to decline in the 1950s and early 1960s, when the novelty of suburban drive-ins lured away its audience. And when the local economy fell flat in the late 1960s and early 1970s, the theater was used as a flea market and then boarded up. Before long it became known as a refuge for drug dealers and junkies.

A turning point for the theater and the barrio came in 1977, when a redistricting measure helped place more Mexican-Americans on San Antonio's city council. Until then, city funding for the arts went almost exclusively to the symphony and a few mainstream art museums. That began to change under the new council, and soon Latino arts organizations were receiving $750,000 each year.

Instrumental in generating support for these grants was the Performance Artists Nucleus, a coalition of Chicano arts groups that merged in 1982 to form the Guadalupe Cultural Arts Center. Its aim was to promote, present, and preserve the Chicano arts in San Antonio.

Troubled at first with infighting and legal problems, the center has matured rapidly since the arrival of a new executive director, Pedro Rodriguez, in 1983. Rodriguez was familiar with the community, having been raised in the neighborhood, and he knew something about art, too. A successful painter and academic, he held tenure at Washington State University as a professor of art history and director of the Chicano studies program. But for Rodriguez, now 56, academia was not enough.

"I felt really isolated up there in the proverbial ivory tower and I wanted to live in a Chicano community again," he says. "I also felt that the mainstream artistic institutions weren't serving the Latino artistic community and that we had to create our own institutions."

Upon returning to Texas, he served three years as director of cultural arts programs for the city of Austin. Then came the chance to head the Guadalupe Cultural Arts Center. Accepting the job, he has not only found his calling but also left his mark.

During Rodriguez's tenure, the center has grown many times over. It now employs a full-time staff of 22, with volunteers numbering in the hundreds, and its operating budget has more than tripled, from $500,000 in 1983 to $1.6 million last year. Its programs in theater, dance, music, literature, the visual arts and film have received numerous grants from the National Endowment for the Arts as well as sizable contributions from private foundations and local businesses.

SOURCE: Reardon 1993, 3-5. Reprinted with permission from *Ford Foundation Newsletter.*

Case Study 7.1. Community Building Through Ownership

Chesterfield has had an impressive recent record of innovation in urban revitalization. It was one of the first cities to respond to the need for rehabilitated housing and to recognize the opportunity of abandoned, vacant city-owned houses by developing the urban homesteading program. City residents willing to rehabilitate and live in the buildings could buy them for one dollar. Chesterfield has become a city that planners, urban government officials, and private developers use as an example of a city willing and able to develop innovative programs to meet the needs of all its residents.

Part of the reason for Chesterfield's success has been the city's ability to build on the strength of its neighborhoods. That process has involved a willingness to listen to neighborhood concerns and to adapt city policies and programs to meet those concerns. Keeping the neighborhood shopping areas alive and healthy has been an important part of that effort.

One of the problems with many city programs is that they attack only one facet of a problem. Cities focus on putting in street improvements or a mall and then hope for the store owners to do their part. They expect homeowners and apartment owners to improve their properties. The city of Chesterfield recognized that public improvements would not, by themselves, have any long-term effect unless there was a total approach to the needs of the shopping areas and homeowners.

The city's staff worked closely with a locally developed City Economic Development Task Force to identify the commercial and housing needs of an area and then organize programs of activities to meet those needs. The program included four major components: public improvements, housing and storefront improvements, management, and financing.

The city's task force recommended a program in which commercial buildings, multiple-dwellings, and a few single family units used as "crackpads" were seized for payment of delinquent taxes or fines. These buildings are being recycled for community use. In the program, as designed by the task force, the city pays for mutually agreed-on public improvements such as public housing, police protection, parking, landscaping, benches, and other street furniture in a designated neighborhood. The citizens of the selected area must work for a nonprofit community-based Community Land Trust (CLT). The CLT purchases distressed properties from the city at cost. CLT properties are then placed in one of three programs–The Lease Purchase Program, the Homeward Bound Program, or the Multifamily Program. Under the CLT Lease Purchase Program, qualified low-income working families who cannot raise the down payment but can make lease payments of $200-$300 per month can lease a home or apartment with an option to purchase. Families must put at least $2,000 in sweat equity into the home on repairs before they can move into the unit. The low-income prospective owner must present a plan to rehabilitate the property at his or her expense. During the lease period the family must keep up the property or face lease termination.

The Homeward Bound Program is aimed at young people who have moved away from their neighborhoods. It is similar to the Lease Purchase Program but aimed at higher income families who still cannot afford a down payment. These families take on mortgage payments of $1,000 to $1,400 per month. Families purchase the property with a low down payment and a second mortgage. They can refinance the project in as early as 3 years and take on the full first mortgage themselves. CLT maintains a ground lease on these properties that specifies owner occupancy and sets a formula for a sales price. If the property is sold for a price higher than the agreed-on formula, the land lease is terminated and the back land rent is due and payable in full.

Finally, the Multifamily Program provides for the rehabilitation of multifamily units by CLT for rental. CLT has a separate housing management company to service the nearly 400 apartment units it currently operates.

A fourth critical area of the Chesterfield program is management assistance. The city has the staff and expertise to do all of the things needed to manage property. The city's program provides that expertise to building owners too small to afford it on their own. The city offers a neighborhood development manager in each district who will carry out an analysis of retail demand in the neighborhood so merchants can offer the goods and services their customers want. They also help merchants plan and carry out promotional events such as street fairs, tours, and market days, and they help with advertising campaigns.

The city is contemplating moving beyond this program to revitalize neighborhood areas. The new approach would be to establish a local community development bank to help finance community- or neighborhood-based projects. The development bank would be formed as a community development bank trust with a local board composed of both neighborhood and business representatives. The city economic development staff that proposed the idea is suggesting that the city council place a portion of the city employee pension fund assets in this bank and the income stream from downtown redevelopment projects as well as the $1 million private sector commitment into the bank. The proposal has generated heated debate in the city council and the community. *The*

Chesterfield Tribune has editorialized that such a use of city pension and investment funds in this manner would be "totally irresponsible." What are the pros and cons for this type of investment approach? Should the city risk its pension funds and city redevelopment income in such projects?

Case Study 7.2. Revitalizing the Center

The key to a city downtown revitalization strategy is a dynamic public/ private partnership. During the late 1960s, Centerville began to decline economically. As in many other moderate-sized, industrial American cities, Centerville's retail sector moved to the suburbs and its industrial base shrank. The 1990 census showed a continuing decline in the city population from 690,000 in 1960 to 499,000 three decades later. Meanwhile, the metropolitan area's population increased from 792,000 to 1,570,000. In the late 1960s and early 1970s, Centerville realized that a dramatic move had to be made to revive the city. That move was a strategy to invest significant public funds to encourage private investment in the city's decaying downtown. Those funds upgraded public facilities and improved sites, opening up new development opportunities for private investors. Spurred by public support, a series of projects have led to a fairly significant comeback for Centerville's downtown. Centerville's success is a useful model for other cities of moderate size in the midst of revitalizing their downtowns.

As a medium-sized city, Centerville had a close-knit and generally progressive business community. The business leaders knew and trusted each other. Those relationships helped to convince them to take some risks to revitalize their downtown. The camaraderie was so strong that once the process began, no one wanted to be left out. That feeling was essential. The city leaders felt that the redevelopment would never have happened if any one participant had failed to join—that is, if the public/ private partnership were less than complete.

In Centerville, 90% of all deals involving public commitments include the city's economic development corporation, Civic Central Area, Inc., a private, nonprofit business organization. The economic development corporation is headed by an executive director and employs 32 people who work to attract new development, retain business, assist small businesses, and package financing. This active municipal component, along with Centerville's mayor and Civic Central Area, Inc., initiated and promoted the downtown revitalization strategy. The public commitment to development has been consistent and predictable, thus making the development environment more hospitable. Centerville's strategy was to initiate development through bold steps that would ripple through the downtown.

The city's early redevelopment efforts were convincing evidence of the need for this direction. One of the first projects, conceived in the late 1960s and constructed in the early 1970s, was a six-block pedestrian mall on Fourth Avenue, downtown. When the mall was on the drawing boards, the retail sector had not completely deteriorated but the suburban shopping trend was clear. By the time the mall was completed, a second suburban shopping center had opened and the true impact on downtown was felt. The mall, unfortunately, was an example of too little, too late. That experience, however, taught Centerville a lesson: only a large investment would affect the downtown's economy.

Centerville's size is also a plus. Although the city has suffered economic decline, as have many Midwestern industrial cities, Centerville does not have the larger pockets of poverty characteristic of larger cities. The extent of poverty, housing problems, and economic decline is not so extreme, and solutions are easier to find.

Yet, the citizen's attitude toward the downtown in a moderate-sized city such as Centerville is harder to break than in a larger city. In Centerville, the public saw no possibility to revive the downtown. In large cities, the central business district frequently has kept its traditional office function, making retail core revitalization easier, but in medium-sized cities such as this one, the public often perceives the downtown as dead for all purposes. Therefore, Centerville launched a major public relations campaign to overcome this negative attitude.

To formulate a new strategy, the city and Civic Central Area, Inc., formed a committee to study the downtown's problems. The city, at that time, had a strong governmental complex at its western edge, a large medical complex on the east side, and the Belvedere Hotel on the north. Fourth Avenue, the heart of the CBD (Central Business District), was clearly suffering.

The committee recommended the construction of the Pigeon Dome, a new center city sports complex, combined with retailing to be located in the heart of the downtown, in order to fill the hole in the donut. The city provided the seed money to plan the Pigeon Dome. The plan envisioned an integrated project of an office, retail, restaurant, and hotel attached to the 25,000-seat indoor sports arena. Municipal support for the project was reinforced with a pledge from the business community to assist in developing necessary feasibility studies and assistance in land assembly. With that commitment, the committee was entertaining a deal with the team owner Al Davids. Al wanted the community to finance the entire project and no rent for the first 10 years. In addition, Al asked for the Pigeons basketball team to control the revenues for all of the stadium parking for the life of their 20-year lease.

The total project cost, $142.5 million, would be financed by the city and state, federal grants, and private sources. Federal funds would be used to acquire land and construct a department store, leased to its operator. City funds constructed the garage from the land monies and built the public open space, including the Pigeon skyboxes. The organized financial and public support from downtown business owners, the city leadership, and the state made a strong partnership—and, the codeveloper believed, an essential one—to make the project work.

However, very significant community opposition has developed. Neighborhood activists have attacked the project as a giveaway to wealthy sports owners. Schools and community groups claim that $150 million could go a long way toward restoring all of the programs cut out of the school budget, lowering class size, and providing housing for the homeless.

Proponents for the Pigeons argue the dome is an investment and cannot be compared with funding the schools or homeless projects. Further, bringing the Pigeons to downtown will prove that a major project in a city of this size would give high visibility all over the Midwest. The project is demanding public attention and dominates the media. The value of the Pigeons playing downtown will bring millions in unpaid advertising to the city. This visibility is a major component in an overall strategy for cities of moderate size, particularly in overcoming negative attitudes and replacing them with civic pride in the downtown.

The mayor of Centerville argues that physical redevelopment must precede business and jobs. An important way to attract new firms is to show potential companies an exciting city with a major sports franchise in its heart. The private and public sectors, working together, can make this happen.

African-American leaders have been especially critical of the downtown Pigeon Dome proposal. They point out that other than a few star players, few African-Americans will benefit from the arena. They have come to the city with a plan that would require that African-Americans be given opportunities for equity participation in all future downtown projects. The proposal before the city council would require that 20% of the equity in the Pigeon Dome be set aside for local African-American investors. Half of this 20%, or 10%, would be from city funds invested in the projects and passed on to a coalition of African-American leaders to be held in a trust fund for minority education. The remaining percentage would be allocated by the developer to African-American residents who would either provide services, cash, or secured notes to gain their share of the project. Finally, the coalition is requiring Al Davids to donate 5% of the team's ownership stakes to a minority development corporation.

Assume you are on the mayor's chief economic development advisory board and develop a winning package for the mayor. Racial overtones could make this nasty.

Case Study 7.3. The Portside Plan

One of the newest trends for raising capital in cities is a "Minority Equity Participation Program" that will increase the participation of minorities and women in new downtown developments. The plan is designed to help pay for affordable housing, economic development, and other needs in city neighborhoods and among minorities. Communities all across the nation from San Francisco to Boston, from Seattle to Jersey City, are attempting to find ways to bring minorities and neighborhoods into the development process. The concept is to design a system that provides for a combination of taxes and incentives to make developers involve minorities in city development projects.

The Portside Minority Equity Participation Program is an innovative "bottom-up" strategy that uses downtown development to fuel growth in the city's neighborhoods. The Portside Minority Equity Participation Program is essential to a city's health in that it can help create self-reliant, locally based entrepreneurship in low-income, minority neighborhoods. Such programs stress self-employment, small businesses, and worker or community ownership emphasizing adequate wages and benefits, decent working conditions, stability, job control, and opportunities for upward mobility. A city is thus rebuilt neighborhood by neighborhood, as small, locally controlled enterprises create jobs for nearby residents.

Portside is giving a whole new meaning to bottom-up development programs by earmarking downtown development—through its Minority Equity Participation Program. The program has three components. First, a new Minority/Community Equity Participation Commission is being appointed. The new commission will have responsibility for establishing guidelines for minimum minority/female participation in all projects with city investments of more than $100,000. The commission is to establish a point system as part of the bidding process so that developers can bid for projects with varying amounts of minority participation, jobs, and other benefits in any project. The project with the greatest overall city benefit would be awarded the opportunity to develop the specified site. Second, a Minority Equity Trust fund is being established. This trust fund is designed as an alternative to direct equity or jobs participation. Developers can contribute to the trust fund in lieu of or in addition to meeting other community goals. Finally, the Minority/Community Equity Participation Commission was authorized to recommend a downtown development fee (tax) on all downtown developments on a per-square-foot basis. The fees collected would be designated for the trust fund.

Within this framework, the Minority/Community Equity Participation (MCEP) Advisory Commission has proposed that a linked development requirement be instituted for downtown commercial development. This would require the developer to design a companion project in one of several low-income neighborhoods. Innovative programs proposed by the committee included a variety of loans for start-up enterprises headed by low-income people, minorities, and women as well as housing and job training projects. One loan program, for example, would provide these businesses with working capital at low interest rates and favorable repayment terms.

Another component of the MCEP Commission's plan was first- source hiring agreements to ensure that the city's most needy residents benefit from the jobs created by downtown development. Any employer who receives a publicly subsidized grant or loan or who is involved in a commercial project involving city financial participation would have to use the mayor's Office of Employment and Training as a "first source" of employment referrals for any new jobs created.

A third component of the MCEP commission's recommendations is a plan for Portside's city government to make half its purchases from city small businesses and 25% from businesses owned by minorities and women.

The Minority Equity Participation Program was adopted without a single dissenting vote of the city council. However, now that the commission is in operation, the city attorney and business groups are raising questions about the legality and the advisability of the MCEP's proposals. The city attorney is arguing that the point system is a violation of the *Croson v. City of Richmond* decision of the U.S. Supreme Court, which prohibits any plan based solely on race. Moreover, the commission, City Attorney Jane Gonzales argues, has not established a nexus between downtown development and the proposed development fee. This, she states, is a clear violation of the *Nollan v. California* decision, which requires a nexus between any tax or fee and the actual project impacts.

The chamber of commerce is livid because the use tax and exaction fees are perceived as antidevelopment. The chamber argues that the office market is too soft to withstand these taxes and that the result of their implementation would be a rapid acceleration of suburban office construction and job growth. "We have talked to developers and those who have done major rehabs, and they say that an additional $2 a square foot [the average exaction fee] can kill a deal," asserts the president of a local business organization.

Several factors have prevented implementation of most of the Portside MCEP proposals. These have included city council fear, developer and real estate opposition, and the unsettled legal questions. The mayor thinks that the city attorney's objections are correct but that her approach is incorrect. What the mayor and a majority of councilmembers want is to find an acceptable formula for implementing the MCEP commission recommendations. Can you figure out a way to get the mayor, the commission, and the city attorney off the hook?

Summary and Conclusion

Locality development is the most well-known technique for stimulating economic development. City officials are generally comfortable with most aspects of this concept. (See Table 7.2 for a

Table 7.2   Locality Development Project Idea Starters

Tool	Project Idea Starter
Community services	*Community radio:* A public broadcasting FM radio station is established to provide independent subscribers access to radio service and to promote an understanding of unemployment and other social disadvantages, together with an opportunity for community groups to give voice to their respective activities.
Community asset management	*Sale and leaseback:* An association sells development rights for public land to an investor who agrees to construct a building (e.g., library or administration center) of the association's choosing, which the association agrees to lease back for several years for an amount equivalent to the cost of the building itself. The investor gains the benefits of depreciation allowances and other tax write-offs. *Computer information service:* This project aims to establish a research and resource center on the use of microcomputers as tools for community organizing. In the first stage of its operation, the center could begin to assess the information needs of selected community groups and compile an inventory of the resources that these groups are willing to share. By entering this information on a database, the center will make it accessible to others with similar aims and needs. Government information could be collected to fill revealed information gaps. The center could also establish computer cooperatives in which groups with similar objectives share the use of computers, enabling them to have resources that alone they would be unable to afford. An electronic calendar to record community group activities and meetings could also be developed to permit groups to coordinate events.

sampling of locality development projects.) Everyone wants to live in an orderly, well-maintained community proud of itself. This does not happen by chance. Farsighted leadership combined with the prudent use of incentives and disincentives can produce the desired results. No community can afford to "sit on its assets."

Table 7.2 Continued

Tool	Project Idea Starter
Household services and housing	*Housing for senior citizens:* A nonprofit organization builds houses in an estate for purchase at a moderate cash sum (e.g., $20,000) by aged persons who agree to deed the house to the organization. *Houses and jobs for youth/low income people:* This has the aim of overcoming housing shortages for youth or low-income people by building housing for medium- to long-term accommodation for this clientele. The construction could provide employment for these persons as well, at least by using unemployed people as builders' assistants under the supervision of local builder(s). Other housing might cater to low-income families.

References and Suggested Reading

Blakely, Edward J. 1992. This City's Not for Burning. *Natural Hazards Journal* 16 (6).

Day, P., and D. Perkins. 1984. Carrot or the Stick? The Incentives in Development Control. *Urban Policy and Research* 2(3).

Farley, Josh, and Norm Glickman. 1986. R&D as an Economic Development Strategy. *Journal of the American Planning Association* 77(1).

Farr, Cherl. 1984. *Shaping the Local Economy*. Washington, DC: International City Management Association.

Frieden, B., and L. Sagalyn. 1989. *Downtown Inc: How America Rebuilds Cities*. Cambridge, MA: MIT Press.

Haar, C., and J. Kayden. 1989. *Zoning and the American Dream*. Chicago: Planners Press.

Hester, Randy. 1985. 12 Steps to Community Development. *Landscape Architecture*, January/February.

Homets, J. 1992. *Small Town and Rural Newsletter* 11 (1/2): 7.

Neutze, Max. 1984. Land Use Planning and Local Economic Development. *Plan*, March.

Oke, Graham. 1986. Targeting Economic Development. *Australian Urban Studies* 13(4).

Purcell, Amelia. 1982. Shopsteading: New Business in Old Neighborhoods. *Commentary*, Spring.

Reardon, Christopher. 1993. Unconquered Spirit. *Ford Foundation Newsletter* 24 (2): 3-5.

Sagalyn, Lynn. 1990. Exploring the Improbable: Local Redevelopment in the Wake of Federal Cutbacks. *Journal of the American Planning Association* 56(4).

Sagalyn, Lynn. 1992. Small Town Newsletter. *Journal of the American Planning Association* 11 (1/2): 7.

Schwab, Jim. 1989. Riverfront Gamblers. *Planning Magazine*, No. 12, September.

Walsh, T. 1992. Dress Up for Company *Economic Development Digest* 1 (9): 1.

8 | Business Development

Business development is an important component of local economic planning because the attraction, creation, or retention of business activities is the best way to establish or maintain a healthy local economy. Such an economy is not merely one that is growing. Community business development is a new entrepreneurial role for planners because, as Frieden says:

> Planners have many compelling reasons for taking on the function of public sector developers. For professionals whose usual complaint is lack of influence, public sector development offers a rare opportunity to get major projects done. For professionals whose usual output consists of planning documents, it provides visible, measurable accomplishment on the ground. And for professionals usually required to have the time-sense of a geologist, it offers the satisfactions of making an impact within a few years.
>
> Getting the projects done poses challenges that many people find exciting. The public sector developer has to mediate between public and private interests in order to find solutions that meet both an economic and a political bottom line. Achieving success means delivering a project that is economically sound and that serves public purposes beyond financial returns. For a planner who wants to influence the character of a city on a large scale, and wants to do it reasonably early in his or her career, public sector development is a promising choice. (Frieden 1990, 427)

Business development is intended to redress the balance between community as a social construct and business as an instrument of wealth generation for planners. As Lockhart puts it:

> community [is] more than a bedroom and service annex to alien commercial interests. Such a notion of community begins with the

178

recognition of the crucial role that the building of shared commitments to the common well-being plays in the attainment of social
health and individual satisfaction. (Lockhart 1987, 57)

In this context, community and business development are merged
as a vehicle to mobilize essential community resources for the generation of shared wealth, both in terms of individual and collective
well-being and in terms of a stronger set of economic institutions that
can compete both locally and globally.

There are at least 12 basic tools or techniques normally considered central to business development. They include the following:
one-stop centers, start-up and venture financing companies, small
business development centers, women's enterprises, group marketing systems, promotion and tourism programs, research and
development programs, incubation centers, micro-enterprises,
technology and business parks, enterprise zones, and entrepreneur
development courses.

The choice of tools depends on the local business development
strategy. Business development strategies can have one or more of
four basic dimensions: to encourage new business start-ups, to
attract new firms to the area, to sustain and expand existing
businesses in the area, and to increase innovation and entrepreneurship within the community. Table 8.1 illustrates how the tools
emphasize the different dimensions. Clearly, they can be mixed
together to incorporate several dimensions depending on the circumstances. In fact, few communities use one tool alone. Most
communities use a combination of tools integrated into the total
local or community economic development strategy.

Creating a Good Business Climate

As mentioned earlier, local governments and neighborhoods can
create a "climate" conducive to business development. Published business climate studies are a mini-industry. Business climate reports on
states, cities and metropolitan regions are published in leading business
and professional magazines worldwide. The significance of these studies remains clouded. After examining their methods and uses, Professor
Rodney Erickson (1987, 62) of Penn State concludes in his study of
business climate appearing in *Economic Development Quarterly*:

There are clearly shortcomings of business climate studies. . . . Does
that mean that we should forget about competitive position and

Table 8.1 Matching Business Development Tools and Objectives

	Objective			
Tool	Business Start-Ups	Business Attraction	Business Expansion/ Retention	Nurturing Innovation & Entrepreneurship
One-stop center	X	X	X	
Start-up and venture financing company	X	X	X	
Small business assistance center	X		X	X
Group marketing system	X		X	X
Promotion and tourism programming	X	X	X	
Research and development				X
Incubation center	X			X
Technology and business park	X	X	X	
Enterprise zone	X	X	X	
Entrepreneurship development activity	X			X
Women's enterprise	X			X
Micro-enterprise	X		X	X

> business climate studies? . . . In my mind the answer is "no." Whether we like it or not, business climate studies are here to stay. . . . [G]ood business climate studies can help to focus attention on the particular nature of problem issues and lead to some worthwhile introspection.

Business climate studies receive media attention externally and internally. Citizens feel better about a well-ranked community irrespective of whether there is no real difference in the economic or social outcomes associated with the subjective measure used in such studies. Most business climate studies measure the comparative aspects of conducting business in one community over another. However, issues in firm location or relocation have more to do with the comparative advantages a city offers for the particular enterprise, such as the nearby location of important suppliers, customers, or scientific information. The key to good climate is in determining what kinds of regulatory and policy tools will facilitate business development for the type of firms that use the locality's asset base, such as a harbor, university, or supplier. This is not an easy task, but it can be achieved. There is no strict mixture with respect to the kinds of incentives and support programs offered. The main goal is to fit the program to the businesses desired and to be flexible.

Therefore, both local neighborhoods and cities need to look at existing regulations to see if they "guide development" to the places and types of activities desired or merely prevent those activities considered undesirable. Finally, the issue of environmental outcomes must be considered as well. The fact that a community has space or facilities for some environmentally contaminating uses does not mean that these should be the best or only industries a community attempts to attract.

One-Stop Centers

One-stop business centers are relatively common components of local government today. In many cases the one-stop center has become a small business ombudsman function located in the office of the mayor, the city manager, or economic development.

The traditional role for a one-stop center is an information center designed to serve as a key contact point between businesses of all kinds and local government. To be a one-stop advice service for businesses, the center must contain information on all planning and development matters of interest and concern to businesses— local economic indicators and labor market statistics, local development plans, land availability, building regulations and permits, all aspects of finance, and other useful business information. It must also be able to make this information available promptly. Clearly, the center's size will be a function of the size and complexity of the economic zone it serves. In large cities and towns, for example, a one-stop center for businesses might employ several people. In neighborhoods or community-level and rural places, in contrast, it might employ one person on a full-time or even on a part-time basis.

A one-stop center is a valuable business development tool. It benefits businesses by eliminating frustrating referrals from one department to another and by saving time that would have been spent, perhaps without result, trying to procure information on local regulations on their own. In addition to assisting businesses, a one-stop center provides valuable services to local government. Statistics from the center can be used as the basis for reports that describe the number, nature, and geographical breakdown of all business requests. The center's data can also indicate short-term business trends or identify potential business development problems. Moreover, the establishment of an organized professional approach to information dissemination for enterprises strengthens business confidence, as well as that of local governments and neighborhoods.

Start-Up and Venture Financing
Companies and Development Banks

Small firms constitute the vast majority of all business enterprises, and they provide employment for the majority of Americans. The formation of new small businesses has effectively utilized both capital and human resources, resulting in employment growth. However, few financial institutions specialize in meeting the needs of small enterprises.

A small business is usually formed using venture capital from the entrepreneur and his or her relatives and friends. If the business is successful, growth is rapid, and the available funds are quickly consumed in buying equipment and/or stock. At this stage, growing small firms need venture capital that will both improve their profitability and lower their risk of failure. The small business owner then usually approaches banks for finance. However, the existing financial system is not equipped to provide adequate venture capital to stimulate the growth of innovative small-size enterprises. Reasons for this shortcoming include constraints on different classes of institutions, an inability of existing financial institutions to evaluate the asset value of highly technical or market-driven innovation, and a general insistence on physical assets as collateral for loans. Banks are primarily lenders to, rather than investors in, small business. Their emphasis on financial security requires some small firms with an inadequate security or equity base but sound profit expectations to seek either higher-cost sources of finance or to defer or cancel development plans if such alternative sources are unavailable. Venture capital is available primarily from private investors or stock issues. However, many projects and business start-ups in the retail or community-service sectors cannot raise funds via the stock exchange, and there are seldom sufficient active local private investors.

The establishment of a mechanism that allows local people to invest in local businesses is therefore an important business development initiative. This can be done through the formation of a

start-up and venture financing company, specifically, a Local Investment Company (LIC). This business development tool can
provide venture capital to selected eligible small firms unable to
obtain finances from other traditional lending institutions and thereby
contribute to local economic and employment development. The
Northern California Loan Fund (NCLF) is an illustration of this form
of investment organization. Headquartered in San Francisco, the
NCLF provides small start-up loans to emerging enterprises in
disadvantaged and rural communities. The NCLF obtains its loan
funds from private investors who agree to invest in the fund and
forgo the interest return on their funds and from foundations and
private corporations interested in socially responsible investments.

Development banks are a larger and more established mechanism
to provide capital for low-income, female, and minority enterprises. A development bank is a regular lending institution that
devotes its resources to fostering economic development (Parzen
and Kieschnick 1992). A development bank might take the form of
a savings and loan, a bank, or a community credit union. There is
now an association of 40 institutions forming the National Association of Community Development Loan Funds. Massachusetts, New
York, and Wisconsin have statewide programs with very similar
orientations, and the Ford Foundation's Local Initiative Support
Corporation (LISC) provides loan funds or capital support for
community development or community enterprise funds on a
regional basis for some community development corporations. By
providing venture capital, local development groups become share-
holders in the companies in which they invest. A community shares
with other stockholders in the success (or failure) of investments;
it is therefore vitally concerned with their long-term development.
Groups such as the Northern California Loan Fund provide the
opportunity for residents within a community to invest in small
business and share in its success, without the risk of investing in
just a single enterprise.

To be successful, an LIC or other community venture group must
invest only in those small businesses, existing or new, that are commercially viable and have the greatest potential. LICs or similar groups
should not invest in businesses that are not commercially viable, even
for social reasons. The community investment program must be run
by a board of directors made up of commercially aware and experienced businesspeople, with relevant business expertise.

In sum, community-based (LIC) groups should prefer to invest in
small businesses that:

- Have an innovation in product, process, or marketing
- Have potential for rapid growth
- Have potential for future sales outside the region
- Can demonstrate sound management skills or, if those are absent, are willing to put them in place
- Are willing, if selected, to establish, with the help of the community, an experienced board of directors to assist in planning the future growth of the business (the composition of this board should complement the skills of the small business owner in finance, marketing, and management)

By using these criteria, a good community venture group can embrace a wide range of innovations or technologies in traditional business areas or in the newer high-technology enterprises. An LIC, in particular, should seek to support a business, not run it, by holding a share of up to 50%. Moreover, because an investment is worthwhile only if at some stage the investor can get his money back, an LIC must have a method of exit via one of the following:

- Selling its share back to the entrepreneur
- Selling its share to another private investor
- Selling its share through the stock market

Finance, particularly local finance that can stimulate new creative enterprises, is the core of small business development. Therefore, every city or neighborhood that wants to pursue this avenue of business development must be prepared at some level to put its money on the line.

Small Business Development Centers

Despite the importance of small businesses as employers and sources of entrepreneurial drive, research evidence suggests that the failure (as distinct from discontinuance) rate among new and small firms is substantial. Poor management is the most frequently cited reason for this. An obvious, and arguably cost-effective, means of improving the economic performance of the small business sector is to establish small business development centers.

These centers provide management training, counseling/consulting, and research services to small firms, with training as the dominant activity. A small business development center should support the training function in various ways. The center that provides these activities could effectively:

1. Encourage a higher rate of new business starts with the potential to succeed
2. Reduce the level of business failures
3. Improve the general financial performance and growth rate of the small firm sector
4. Raise the potential of small firms to create new jobs and improve employment levels
5. Raise the general level of technological innovation and productivity

Although small business training/research/counseling activities may be provided through any of several alternative organizations, the best form is a specialist institution organized in close association with, rather than within, appropriate educational institutions (such as university business schools, advanced education, and technical colleges).

Another approach is to turn unemployed workers into business owners and operators through a business resource center. The purpose of these centers is to tap the entrepreneurial talents of laid-off workers and find niches in the local market where they might start their own enterprises. Business resource centers are usually started by local service clubs, the chamber of commerce, local employment and training institutes, or even a separate nonprofit development and training company. This type of nonprofit organization is designed to use the skills of the underemployed as a basis for generating new employment using business resource centers. These centers provide the following:

- Practical training in business start ups
- Low-cost, small premises
- Centralized services such as photocopying, telephone answering, and accounting
- A big brother or sister for the new business

Group Marketing Systems

Group marketing occurs when several self-employed individuals or companies come together to share some or all of their marketing and distribution activities. The main economic factors that encourage the establishment of group marketing operations are these:

- Limited economies of scale in production
- Large economies of scale in marketing and distribution
- The need to survive a common external threat such as high-import penetration or a fear of closure due to industrial decline
- Positive attitudes toward business collaboration

In industries and localities where these economic factors prevail, joint marketing ventures are likely to prove successful and rewarding to the participants. Group marketing provides many benefits, most important being the improvement in overall standard of competence and product quality of the participating firms. Group marketing is most successful when it is limited to achieving specific objectives and includes a very small group of participating firms that share the same market. Further, there must be easily perceived real returns for participants in group marketing schemes. The members must also agree on the financial commitment and the time scale and trust one another to keep their commitments. Finally, government involvement at the local level is an asset as well since local governments can actively encourage these arrangements and market them to prospective customers.

Potential barriers to group marketing include:

- Lack of awareness of the need to improve marketing or of the benefits of a group approach
- Reluctance to change established marketing methods
- Preference to remain small, especially in craft activities
- Fear of losing control over planning and management
- Jealousy and mistrust of other enterprises
- Lack of finance to promote new schemes
- Legal constraints such as inappropriate legal frameworks for developing cooperative marketing enterprises except for agricultural cooperatives

Local governments and other institutions that wish to promote the use of this business development tool should attempt to overcome the barriers described above and put the right people together. Some additional tips for promoting joint marketing initiatives are to do the following:

- Talk to the principals in companies, not only their representatives who might not report back to their colleagues
- Design any proposed venture on the basis of the needs, problems, and opportunities of individual companies identified through interviews, rather than on perceived "common interest"

- Present the full package to prospective collaborating companies, rather than proposals short on detail or full of options
- Make sure that the level of financial commitment is realistic and can be seen to be value for money in relation to the visits involved (it is better to frighten off potential collaborators at the beginning on cost grounds than for them to fall out at a later stage)
- Leave it to the group of companies to select the range of complementary products and to decide what approach to adopt to tackle the market

Group marketing projects tend to take about 3 years to become properly established and are then likely to go from strength to strength. An action plan for group marketing would resemble the following one:

- Identify a group of firms interested in working together; there should be 4 to 12 firms.
- Research individual needs and problems (third-party interviewing).
- Establish common objectives.
- Agree on complementary noncompetitive strategy.
- Select a range of complementary products.
- Define methods of approaching the market by using a joint catalogue or directory or a joint sales agent.
- Research markets and distribution channels and select specific target markets in areas within a market.
- Undertake a feasibility study and ensure that likely financial commitment for consortium members is in place.
- Draw up a legal contract among members specifying the financial commitment.
- Nominate or recruit a co-coordinator.
- Negotiate only with principals of firms.
- Keep deadlines tight.
- Have the package up front for the consortium to respond to.
- Watch out for "pirates" just wanting to take ideas from others.

Women's Enterprises

One of the most underutilized resources in the nation is the entrepreneurial talent of millions of women. This is particularly true of women heads of households. In the mid-1980s, a pioneering organization was established in Minneapolis-St. Paul to assist women of low income and on welfare to start their own businesses. The mission of this organization, called WEDCO, now known as Women Venture, was to assist women developing small enterprises through an intensive self-help program.

Women Venture has more than 10 affiliate operations around the country. Each is based on a very similar model. Low- and moderate-income women are put through a 7-week intensive training program. The program focuses on assisting them to identify their talents and interests and on building the self- confidence to start a competitive enterprise. The program is rigorous, with few dropouts. Although not all women start their own businesses, many complete the training with enough confidence to seek work in challenging and rewarding fields.

Promotion and Tourism Programs

The tourism industry has grown rapidly over recent years. Communities have favored and continue to promote tourism in the belief that it is a significant economic and employment development tool. However, surprisingly little is known about the economic and employment implications of this new growth industry. Economic and social studies of the tourism industry, which are purely objective in nature, remain few in number. Such studies have been hampered by a lack of consistent data, particularly at the regional level. Studies of tourism that measure the impact show that tourism at a national and state level is not a major industry in terms of its contribution to the gross domestic product or employment.

However, these results obscure the vital fact that tourism is spatially selective. Tourism is not ubiquitously but regionally located. As of yet, no comprehensive studies of the local regional significance of tourism have been made. Nonetheless, three types of regions can be recognized with respect to the role of tourism:

Regions lacking tourism significance. These are predominantly rural inland regions with no particular or unique attraction for the public traveling for recreational purposes, and towns off the main business travel routes. They include most of the nation's small towns, particularly those in the Midwest and Southwest. However, even a community without observable tourism significance can alter its destiny by offering an unusual opportunity such as a "weekend on the farm" or capitalizing on a specific historical event such as the birthplace of a famous or infamous person.

Regions of high tourist significance. These are areas with substantial scenic wonders and excellent climate. They have national and international appeal without increased promotion. They are generally in or near large metropolitan areas. Among those that come to mind immediately are such places as Yosemite, San Francisco, or Miami.

Centers of some tourist interest. These are places that have tourism as a component of some other major economic activity. Southern California's film and television industry provides this form of tourism/recreation opportunity.

Although tourism can be an asset, it does not provide the solution for most of the regional/urban areas currently in a state of no-growth or decline and should not be promoted as such a solution. Indeed, available research on regional economics indicates that local demand for goods and services is more important than visitor demand. Even in those instances where the tourist industry has been a major source of growth or decline, there is some doubt that it has ever reached a level of economic importance equal to that of the local permanent population growth. If this is the case, it follows that local planning authorities should lean more toward attracting more permanent residents—retirees, business managers, and the like—than toward seeking to boost tourism from distant places as a source of economic and employment development. Of course, the need for social and physical infrastructure development to keep pace with the rate of population growth must be borne in mind. In sum, for the majority of American communities, tourism is best seen as only one component of an economic development strategy rather than as the entire strategy itself.

To be successful, a tourism program of any kind must be well planned and managed. It must have explicit objectives, such as business development, as well as the active support of all local parties with an interest in and effect on tourism. It should be planned around specific themes such as business conventions for the industries identified in the region's industrial development strategy. It should also be targeted toward specific populations, such as nonlocal businesspeople, and be promoted and marketed in the places of residence of these populations.

Micro-Enterprises

The concept of micro-enterprise is borrowed from Bangladesh. The basic concept is to loan funds to a group of borrowers who plan to go into very small labor-intensive businesses in the same community. Micro-enterprises are usually operated from the home or on the street. They include such enterprises as homemade jewelry or garments, handmade shoes, or specialty foods for restaurants.

Micro-enterprise programs usually make very small loans to a group of 5 to 10 borrowers, each borrowing under $1,000. The group is responsible for each member making his or her loan payments on time. There is usually some form of pre–business training and group building before loans are made to potential entrepreneurs. These programs are now in place in many parts of the country. Although these efforts are too new to assess, there are several in operation. These include Micro Industry Credit Organization in Tucson, the Self-Employment Circle (a subsidiary of the South Shore Bank in Chicago), and the Lakota Fund on the Pine Ridge Reservation in Kyle, South Dakota.

Research and Development

A dominant feature of today's economies is the knowledge-intensive high-technology enterprise. World market competition requires advanced nations to develop value-added and research-intensive products to remain competitive. As a result, technological developments require an increased knowledge base for industry. The microcomputer industry, which both receives and supplies information, is a prime example of the use of intelligence as a product. But many other new information and telecommunication-related technologies exert profound influences on traditional production technologies. Indeed, it can be said that all tomorrow's industries will be knowledge-intensive, or they just will not exist at all.

An economy based on new environmental industries, biotechnology, and information technologies is emerging. Since the first edition of this book in 1989, the cellular telephone business has emerged and now dominates telecommunications marketing. Many of the new start-up industries need a place to start. They require a community interested in mixing uses such as housing and retail. Communities that provide the right atmosphere for start-up firms will attract them. If they nurture these firms by providing the right kind of atmosphere, they can become the headquarters for new companies or the centers of revitalization of older ones.

Local communities are well placed to identify their target technology sectors and activities. Moreover, investments should be directed toward developing infrastructure to suit new firms. Improved links between universities and other research organizations and industrial and business companies should be encouraged and more joint funds established by companies and universities for common research and new product commercialization.

Inventors need a supporting environment during the crucial period between the conception of a new idea and its development to the point where it can be taken to the venture capitalist. Lack of a supporting environment during that critical period can result in many fine ideas never taking off. Incubation centers (also known as "technology development centers") are intended to provide this supporting environment. They provide low-rent workplaces and are usually located adjacent to educational institutions, which provide small inventors with equipment, facilities, advice on business management, and work support from student assistants to put together business plans and to approach venture capitalists. Control Data has pioneered in the establishment of incubation centers. Other centers can be found within some technology and business parks.

Technology and business parks, in this context, refer to parks that house commercial activities rather than research and development activities. They usually cover several acres and have the following general features:

- A campus-style physical environment
- A mixture of ownership and management, ranging from government to private sector
- A frequent association with some form of university or science-oriented establishment
- Low-density development with high-quality buildings
- Specific criteria for the eligibility of prospective occupants to ensure that all the activities within the park are compatible with each other

The technology and business parks' development is in its infancy, so it is difficult to draw any significant conclusions about its likely success. However, ingredients for success apparently include the following:

- Adherence to a carefully thought-out and very specific development objective, including type of desired activities
- Availability of good management and entrepreneurial skills
- Access to venture capital and the accompanying advice on how to use it
- Involvement of the private sector in planning the development
- Presence of meaningful community support

Other locational elements that appear to be influential include transport access, availability of skilled labor, an attractive living environment, proximity to appropriate research facilities, and the presence of a suitable major manufacturing enterprise that acts as a catalyst. These findings are derived from a range of empirical studies on technology parks. It is important to note that these studies commonly suggest that the development times for technology and business parks are long, generally 10 to 20 years.

Enterprise Zones

An enterprise zone is a defined area where planning controls are kept to a minimum and attractive financial incentives are offered to prospective developers and occupants.

The concept of enterprise zones was introduced in the United Kingdom in the late 1970s based on the Hong Kong record of increased job attraction through reduced regulation. They were designed as a last desperate measure to be tried only on a small scale in depressed areas where conventional economic and employment development policies and tools had failed to arrest a decline. An underlying assumption of the concept is that the removal or streamlining of certain statutory and administrative controls would encourage entrepreneurs to create or expand businesses and in the process provide the jobs and environmental improvements that depressed areas so badly need.

To achieve this objective, policymakers need to emphasize labor-intensive industries, and in particular, that subgroup of labor-intensive industries that offers employment opportunities that match the skill levels of potential workers in the distressed area. This can be attempted through strong wage subsidies in addition to tax relief packages and land availability. Thirty-seven states have created enterprise zone legislation. Louisiana has 750 enterprise zones, and virtually all of Toledo and Cleveland have been declared enterprise zones under state legislation. Federal legislation will provide between 50 and 100 enterprise zones during the Clinton administration. Although enterprise zones have many advocates, the jury is still out on their real benefits. Recent studies indicate that enterprise zones do generate jobs because of the concentrated attention of policymakers and not entirely because of the incentives provided by localities or states (Green 1991; Rubin and Wilder 1989).

However, policymakers need to avoid several pitfalls. These include the temptation to view enterprise zones as potential high-technology parks, ignoring any relation between the employment structures of high-technology firms and the skill levels of the local workforce, and the tendency to offer incentives that lead to the relocation of existing businesses rather than the attraction of new ones, thereby influencing their position but not their overall productivity.

Other potential disadvantages of economic zones are that large developers and well-established corporations are likely to be the major beneficiaries. They are likely to place an unfair burden on existing commercial taxpayers outside the zones.

New Entrepreneur Development Activities

An entrepreneur, as defined in the Oxford dictionary, is someone who undertakes a business or enterprise with chance of profit or loss. Put more simply, an entrepreneur is a risk taker. Not all risk takers, however, are successful. To be successful, the entrepreneur needs to be proficient in all aspects of decision making. Successful decision making by entrepreneurs involves three steps:

1. Self-knowledge, or the knowledge to specify the objective; imagination and analytical ability to focus on the factors that lead to decisive action
2. Research, for which search skills are necessary for collection of data and foresight to estimate data
3. Decision making and implementation, which require computational skills for applying the data to the decision rule and communication skills for formulating implementation plans

In addition to these skills, the entrepreneur needs to have delegation and organizational skills to involve and utilize specialists as required.

Cities can facilitate the identification of people with entrepreneurial talent and interests and develop them. One of the best methods for doing this is to start youth business programs in the schools and set up entrepreneurial training programs in the junior or community colleges. Some big-city newspapers such as the *Oakland Tribune* in California have sponsored business start-up fairs and training sessions.

Local governments can provide a supporting environment for investors and small businesses through the establishment of incubation centers and business assistance centers. Second, governments can create a positive economic development attitude throughout a community and thereby provide entrepreneurs with an environment in which they can flourish. These efforts all create a supportive environment for risk taking. Although not all entrepreneurial skills can be taught, at least the values of risk taking can be reinforced by community action.

Illustration 8.1. Determination Helped Snatch this Town from the Jaws of Economic Death

Cuba, Mo.–The metallic brown Ford van came barreling down the gravel road, kicking clouds of dust toward the overcast sky, its horn beating out a rhythmic warning sound.

"Oh, my gracious," said Dennis Roedemeier, president of the Industrial Development Board of this little Missouri town, "something's gone wrong."

As the van came skidding to a halt, grim-faced Mayor Ray Mortimeyer leaned out of the window and said, "We've lost the grant."

"What went wrong?" Roedemeier asked. Even as he asked, he knew nothing had gone wrong, because the mayor could not contain himself any longer and broke into a broad grin, extending his beefy hand in heartfelt congratulations.

Minutes before, the state community development block grant office had notified the mayor that the city had been awarded $900,000 in state grants to enable the Bailey Corporation to open a facility in Cuba's new industrial park. Bailey is the fifth new company to locate in Cuba this year. In addition, two local companies have expanded their operations, and a shoe plant that had closed down has re-opened. These moves have created 450 jobs. Cuba, unlike so many other small towns in rural America that have been victimized by cheap imports and regional reces-

sion, did not roll over and die. It has bounced back from the dark days of the autumn of 1984 when this town of 2,100 people lost 125 jobs and saw its unemployment rate soar to 13 percent.

It took dreaming and scheming, naive faith and dogged persistence for Cuba to turn its fortunes around.

Cuba lies alongside Interstate 44 in south central Missouri, about a third of the way between St. Louis and Springfield. The city advertises itself as "the gateway to the Ozarks"–a testament to the region's beauty, but also its poverty.

Two out of every three people in Cuba live below the state poverty level; until recently, things were getting worse as the area's principal industries–lead and iron mining and shoes–declined in the face of foreign imports. "We're surrounded by small towns that built their future on the one little factory," Mortimeyer said. "As that plant closed its doors, so did hope."

Cuba seemed destined for the same fate . . . But there were people in Cuba who refused to let their town die. Mortimeyer, who had operated a local home heating fuel business since 1969, was elected mayor in April 1984 on a platform promising an activist government that would run Cuba more efficiently.

One of the things the new mayor did was to convene the local Industrial Development Board, which had been in existence since September 1983, but never met. He asked Roedemeier, a local entrepreneur who had recently sold his business, to become the volunteer president of the board. Roedemeier initially demurred but finally relented after assurances that business executives would run the show and politicians would not interfere.

Roedemeier knew that Cuba had to act quickly. "We had been slowly bleeding to death and then we began to hemorrhage," he said. In September, 1984, 55 more jobs were lost when Prismo Paint Company and Echo Shoes closed their doors. In November, the Mid-America Shoe Company announced that it would close by the end of the year, taking with it 70 jobs.

Roedemeier visited neighboring communities. He talked to their business leaders and elected officials to see what they had done right and where they had gone wrong. He prowled the corridors of the state capital in Jefferson City, learning all he could about the myriad of state assistance programs for business development.

That dogged determination paid off. From January to November, 1985, Cuba received $1.6 million in state grants to attract new industry, to help existing companies refurbish, and to provide necessary improvement in public services.

Such financial assistance coupled with a new spirit of entrepreneurial creativity rapidly began to show results. In February, the Mid-America Shoe Company that had shut down re-opened as Whistles, Inc., planning to overcome mounting foreign competition with local commitment and innovation. The plant's former manager and salesman mortgaged their homes and bought out Mid-America's owners.

The city received a state development action grant of $119,000 and lent it to Whistles at a seven percent rate to enable the company to purchase new equipment. A group of concerned local business executives bought the company's buildings and agreed to lease it back for a nominal sum.

SOURCE: Stokes 1985. Copyright © 1985 by National Journal Inc. All rights reserved. Reprinted by permission.

Illustration 8.2. The Bank with a Heart

People tend to use glowing terms when they're describing Chicago's South Shore Bank. Some call it a pioneer. Some say it's a savior. Still others call it a social experiment—not a bank in the conventional sense but an academic version of a bank where social theories take precedence over financial results.

One of the four people who took control of the bank in 1973 and is now its chairman calls this peculiar institution "a handyman." A community development bank, Milton Davis told a U.S. Senate committee in February, is "intimately familiar with particular local problems" and has a toolbox full of gadgets to help fix them. Credit is

one of those tools, but only one. Persistence is equally important, he and his colleagues insist.

Over the past two decades, they have applied those principles to South Shore, a two-square-mile, predominantly black neighborhood (pop. 61,500) with a median household income of $15,909 and a poverty rate of 27.3 percent, located nine miles south of the Loop. In 1970, the neighborhood was a question mark. Its racial mix had switched from mostly white to mostly black in the preceding decade, and local retailers and institutions had begun to disinvest. The area's one remaining bank, then called South Shore National Bank, asked the federal government for permission to leave the neighborhood and move downtown.

Instead, control of the bank shifted to a civic-minded management group—all of them experienced bankers—that has stuck together for 20 years, concocting revitalization strategies that eventually attracted the attention of policy experts nationwide.

Shorebank has grown from $40 million in assets in 1973 to $204 million in 1992. At the end of 1991, it had $188 million in deposits, and $142 million in loans outstanding. Through the efforts of the bank, a development company (City Lands), and a job training and family support service (The Neighborhood Institute), Shorebank has helped renovate 12,500 housing units in five Chicago neighborhoods. In the South Shore neighborhood alone, it has helped rehab more than 7,700 apartments, almost 30 percent of all the area's rental units.

The bank's managers are convinced that their methods can be copied because they've done it themselves—in other Chicago neighborhoods, elsewhere in the U.S., and in Poland. President Bill Clinton has said that he wants to create 100 community development banks modeled after South Shore. Shorebank officials caution, however, that they can't parachute into 100 cities and create clones of South Shore Bank. In fact, they're not convinced that banks are the only antidote for disinvestment. The message is: Watch what we do, but adapt our ideas to your own circumstances.

"Shorebank starts where the conventional market stops," says Joan Shapiro, a senior vice-president of both the holding company and the bank. "We have a double standard. We have to achieve both development and profit results in all our work." In this context, "development" is a code word for "doing good"—a notion that probably gives most bankers the willies.

Some observers argue that, in fact, Shorebank couldn't be profitable without its social focus. "A successful development effort did not necessarily mean the long run sacrifice of profitability," writes Richard Taub in *Community Capitalism,* a history of Shorebank through 1986. "Part of the bank's development ideology includes the belief that as the community becomes more prosperous, so does the bank."

South Shore is the laboratory where Shorebank's theories are tested first. Here's how some of the strategies have worked:

Theory: Fix up the housing, and the commercial areas will follow suit. Results: It hasn't worked out that way. The residential streets right around the bank look healthy, but most of the nearby shopping streets don't.

Theory: Renovate the worst buildings in the neighborhood, and local people will borrow money to rehab smaller, more manageable properties. Results: Of the 7,700 housing units that have been rehabbed in South Shore in the past two decades about 1,400 have been done by City Lands or The Neighborhood Institute. The rest were rehabbed by individual building owners who borrowed from South Shore Bank to get the job done.

"The theory is that if we do the biggest, nastiest buildings in the most prominent locations, it will stabilize the neighborhood to the point that private owners and mom-and-pop rehabbers can justify doing the small buildings," says Lynn Railsback, a vice-president of City Lands.

Theory: Create jobs, and people will have enough money to pay the rent for these newly rehabbed apartments. Results: Shorebank has helped create several hundred new jobs in South Shore since 1990. But the net income of neighborhood residents had actually declined slightly in the previous decade. Median household income was $15,923 in 1980 and $15,909 in 1990 (in 1982 dollars), according to a recent Shorebank report on social indicators. Poverty rates rose from 23 to 27 percent in the same period.

The Neighborhood Institute is helping to make people employable as well as employed, says Linda Green, its executive vice-president. Besides housing rehab, TNI does training, job placement, and job creation. It has opened three incubator buildings in South Shore with about 50 businesses, and it has placed about 200 local residents in jobs at a major supermarket that opened in 1990 in Jeffrey Plaza, a neighborhood shopping strip developed by City Lands and partners.

"If there's one thing that Shorebank has learned over the years," says Green, "it's that we can put up all these pretty buildings, but if we don't deal with the human component, it means virtually nothing."

This approach leads in some surprising directions. In South Shore, TNI has created a program to help fathers become better parents, and it is organizing an anti-crime campaign. In Chicago's Austin neighborhood—where Shorebank set up shop in 1986—City Lands recently opened a library with 4,000 books and tutors for local schoolchildren.

South Shore Bank works hard to make sure its loans don't go sour. It brags about having a loan-loss ratio lower than that of other banks with similar assets—in 1992, around four-tenths of one percent of total dollars loaned. To keep losses to a minimum, Bringley says, he finds new borrowers to take over troubled properties.

Unlike most banks, South Shore Bank makes more real estate loans than commercial loans. At the end of 1992, the bank had $86.3 million in real estate loans outstanding, more than half its total loan dollars. Real estate loans "are the most profitable thing the bank does, and they have the most development impact," Bringley says.

As every borrower knows, banks make a profit on the spread between the interest paid on deposits and the interest charged on loans. Banks that slurp up deposits from a community but refuse to make loans there are guilty of redlining. What South Shore Bank does is greenlining.

SOURCE: Lewis 1993. Reprinted with permission from *Planning*, copyright © April 1993 by the American Planning Association, 1313 E. 60th St., Chicago, IL 60637.

Case Study 8.1. Community-Based Business

"The bankers aren't interested. They tried this in the 1960s, but they had a high rate of loss." The speaker was the head of the new government-guaranteed bank program. Reunion, a community business development bank, had approached local commercial banks seeking participation in a loan program to finance new inner-city ventures. The Reunion Development Bank was making overtures to use the commercial banks' Community Reinvestment funds for community purposes. The commercial banks they were approaching had tried inner-city loan funds in the 1960s and 1970s, and were disappointed by the loan losses.

Commercial, small, and minority business loan funds of the 1960s and 1970s suffered from a number of major shortcomings. These funds often employed weak loan criteria and had little or no oversight; and bank managers showed little concern when the loss rates began to mount. In recent years, however, new community-oriented funds have emerged that have learned from past mistakes. Although the community development banks are a new instrument, they have shown considerably greater success in their repayment rates.

The Reunion Bank's Loan Committee has nine new projects identified across the city which illustrate the success of the community development bank concept. One of the best examples of community-oriented loan programs is the Business Development Corporation of Centerville (BDCC). The emphasis of the Centerville BDC is on Hispanic entrepreneurship to provide jobs for the area's unemployed population. Although capitalized with nearly $1 million in city funds, BDCC's first years were disappointing, in part because the group followed the Small Business Administraion (SBA) model of administration. Community representatives, inexperienced in banking, made decisions. By BDCC's third year in business, the loan loss rate was reported to have reached nearly 60%.

In that year, the city restructured the loan committee to include banking professionals, and a former commercial bank official was brought in as executive director. Then the picture began to change. During a three-year period, BDCC made 32 loans to small businesses, most of them in the range of $10,000 to $50,000, with three- to five-year repayment schemes. At present, only three of the loans are in default.

Four key ingredients make the new loan funds effective. They are:

1. Professional criteria. It is important to use strict criteria in financing decisions. These criteria include the applicant's experience in the field, his or her investment of capital, and a convincing business plan.
2. Careful monitoring. Try to keep a manageable volume of loans so that each one can be closely monitored and identified if management assistance is needed.
3. Training and technical assistance. This complements the basic entrepreneurship and should be offered, and in most instances required, of loan recipients.
4. Involving others. Most of the new loan funds, unlike their predecessors, provide only part of the total loan needed to start a business. For example, the BDCC will provide financing for only up to one-third of the total package, requiring applicants to find other private sources. The involvement of other lenders provides additional analysis of business plans and enables the loan funds to leverage their participation.

The loan funds of the 1960s and 1970s had disappointing results, but the new loan funds have benefited from the mistakes of the past. They can grow into a major antipoverty force only if they can hold true to entrepreneurial values: creativity, decentralization, and market discipline.

The Cadillac Foundation has become interested in minority lending. The Foundation Directors, chaired by John Jonas, have directed the staff to consider forming a loan guarantee plan for community development banks. Cadillac wants to support loans to the homeless through BDCC on a model similar to the women's enterprise programs. How would such a program be designed by the community development bank, or is this just too risky for BDDC? Mr. Jonas is adamant about introducing a micro-loan for the homeless, but how would you devise a program using all of the methods detailed above to move the homeless from their current status to self-sufficiency?

Case Study 8.2. A Small Town Determines Its Own Destiny

Steelnet's dream evaporated almost overnight. The Midwest Iron Range community, located on a two-lane county road in the northern Great Lakes region, began for just one reason: a low-grade iron ore called taconite. The Reservation Mining Company built the town in 1939 expressly for its workers, who earned good wages, financed their houses on easy terms through an employee credit union, and often had money to burn.

For years, Reservation's taconite mine, a joint venture of the Inland and Republic steel companies, was the most productive in the world. But decline in the U.S. steel industry and competition from other taconite sources finally closed the mine, decimating the town that at its peak was home to 30,700 people. Lacking proximity to anything but its iron mines, Steelnet had to scramble for economic survival. So did its residents, many of whom left as unemployment soared to 85%. The rapid exodus cut the population from over 30,000 to today's 12,300. The average housing price plummeted to under $15,000 per single-family home.

Steelnet could have become a ghost town. When the mine reopened a few years later, it recalled just 100 of its 1,500 former workers. What has kept the town alive—and optimistic—is the tenacity of its citizens and the mayor, Don Cole, a retired mine foreman. A smile and an easy laugh are part of the salesman's repertoire, and Cole is a salesman for Steelnet. His first priority was to train the town to market itself through a program, launched by the state's Department of Energy and Economic Development. Named "Star Cities," the program is designed to highlight and strengthen the state's most aggressive local development programs.

The certification requirements to become a Star City would daunt many small cities, but Steelnet had one key asset: a pool of unemployed volunteers who combined their talents to make the town a Star City in a record four and a half months. The achievement won it visibility among state officials, which paid handsome dividends later.

The Steelnet city council also formed the Steelnet Area Development Association (SADA), inviting joint efforts with a nearby township that shares the same school district. Cole and the council never planned to settle for a passive role, even though a small board of citizens assumed control of economic development efforts. The same civic enthusiasm that achieved a speedy Star City designation enlisted 435 individual and 29 business members to sustain SADA's projects and committees.

Mayor Cole joined the search for funding sources. He recalls an effort to solicit foundation officials who had never heard of Steelnet and had no idea where it was or why they should care. Cole was a good salesman. He was able to attract foundation grants that provided $55,000 for the city's revolving loan fund, targeted for small business loans to local people who could not qualify for bank credit. Steelnet's skilled but largely unemployed workforce would become its main source of entrepreneurial talent for these loans.

Because the city had relocated its offices to a retail/office complex in a converted school building, the old city hall became an ideal trial site for SADA's efforts. A community development block grant financed renovation of the building, while SADA loans allowed a pool table manufacturer and a small novelty products firm to occupy it. SADA has relied heavily on board members' personal knowledge of loan applicants in taking risks. Yet, SADA's 20 loans have shown a perfect repayment rate. The city did less well at first, however, in trying to attract new businesses from other areas. On the bright side, Reservation Mining's steady decline offered new opportunities for the reuse of abandoned buildings. A rubber tire recycling plant, located in a building formerly used by a pneumatic drill manufacturer and bought by the city for $5,000, is expected eventually to create 60 new jobs; two firms that will also locate in Steelnet to make use of the rubber may add twice that number. Steelnet's strategies finally seem to have paid off. This little town didn't give up or die. There is a lesson here for many small towns dependent on natural resources or a single industry.

One of the union leaders is opposing the plan for the reuse of the mining facility by an outside firm. He argues that outsiders are nonunion and that these union-busting firms are the cause of Steelnet's problems. Joe Bartowski, a former union steward, argues that "we should loan some of that money to ourselves and reopen the rubber plant ourselves."

"After all, we worked in it before," he says. "Who knows the market and the products better than we do?" Mayor Cole thinks Joe is a little crazy. "Working in a place ain't like running it." Well, who is right? Maybe they both are, for different reasons? Or is there something in what Joe says that can be used by Steelnet before it lets the new firm in?

Case Study 8.3. An Incubator as a Revitalization Tool

When the Kanter Corporation, a metal fabricating company, abandoned a 73,000-square-foot facility on three acres of land one mile north of downtown Newton, Kanter gave the land and buildings to the city for one dollar. The City of Canton originally wanted to use the facility for a new community college, but a feasibility study showed this would not be financially prudent. The Greater Regional Economic Development Foundation then asked the city council to allow it to investigate the site's suitability for an incubator facility.

Canton was able to obtain a Small Cities Block Grant to perform a feasibility study. The study, which included a building and site evaluation, marketing strategy, and incubator plan and program, came back favorable for an incubator facility. The Greater Regional Economic Development Foundation obtained a $70,000 matching grant from the foundation and a $70,000, 5% loan from the city, with repayment beginning in 3 years. In addition, the city charges the incubator no rent. The $70,000 from the city is money that the city would have needed to spend on security, maintenance, and so on, for the incubator building if it had remained empty.

The Women's Equity Development Corporation Center will run a business incubator for women-owned firms in Canton. Women who want to rent or lease space may do so only after they have completed a special training program organized by the Women's Equity Development Corporation (WEDCO). The incubator offers newly established businesses the opportunity to rent space at below market rates and to have access to free telephone receptionist services, secretarial services, business management assistance, custodial and maintenance service, and a business technical library. In addition, the businesses have an opportunity to network with each other. Most of the entrepreneurs at WEDCO are either first-time female-owned business owners or are moving to the incubator from their garages, basements, or barns.

The WEDCO incubator currently houses eight businesses, including a manufacturer of wooden playground equipment, a company that sells bagged ice to convenience stores and bait shops, a firm that makes tools for air conditioner repair, and a computer services business. Lake College also plans to relocate its Business and Industrial Institute to the incubator. The institute will offer 50 different programs, including technical manual writing, computer training, management training, and blueprint reading. Each business in the incubator will be entitled to one free program. The Lake College programs are developing into an entrepreneurial training institute for businesspeople in the area.

The primary function of the incubator's Operating Committee is to screen applicants. More prospective tenants are rejected than are accepted. The major criterion for acceptance is the business's ability to create jobs, either now or in the future, for women. The incubator is not promoting professional people or service sector businesses that offer limited job creation but is instead emphasizing manufacturing and information processing businesses.

The incubator staff assists approved businesses to prepare a business plan and steers them in the direction of local lenders who have been extremely supportive in lending start-up capital to incubator businesses. Each company occupies 500 to 2,000 square feet at a rental rate of $3 per square foot per year. Leases can be for up to 5 years, but no business has signed on for more than 2 years, hoping to move to larger quarters by that time. The incubator is expected to create 120 jobs and currently has a 65% occupancy rate, ahead of the 15% predicted by the end of 1 year of operation.

The community goal of the incubator is to create jobs and ownership for women. Some of the incubator businesses may be able to move to the city industrial park in a few years to help revitalize this city-owned, underutilized facility. By the third year of the incubator's operation, it should be self-sufficient and be able to begin repaying the city loan.

The mayor and the Greater Regional Development Foundation are now in the process of calculating the return on investment. They are unfamiliar with the appropriate methods to use. The building with improvements is valued at $1 million. Most of the rental spaces will rent at lower than market rates, as discussed earlier. A consultant has been hired to assist making the calculations; however, the committee wants to provide the consultant with the parameters for the study before it commences. What are the variables that need to be considered, and how will the city know if it is in fact making both a social and economic profit?

Summary and Conclusion

Business is the heart of economic development. All business originates from a sense of opportunity to serve or gain. Irrespective of the motivation to start a business or maintain it, this motivation can be hampered or retarded by the actions of communities or local governments. No business wants to be where it is not wanted. However, communities need to take extreme care with the type of firms they encourage. Linking, as a community, economic and employment objectives to the types of firms in the locality is very important. (See Table 8.2 for a sampling of business development projects.)

The means communities use to attract firms will depend on what resources are available. The most important factor in firm location is knowing the area's assets as well as having the correct attitudes and infrastructure to support the firm. Every community must develop special activities and tools to build the right type of infrastructure for the development and the employment it wants. Chasing tourists or high-tech firms should not top the agenda unless the community is particularly well-situated for this form of economic development. The most important activity is building soft infrastructure, such as information and finance, to make the community attractive to enterprise. In essence, the community must become an entrepreneur in the use of its resources if it wants

Table 8.2   Business Development Project Starters

Tool	Project Idea Starter
Attraction	*Organization development:* A local development commission with state finance backing loans funds to a nonlocal dying industry suited to the region to assist that industry to relocate and develop as a business development venture and possibly also as a tourist attraction (e.g., a local development commission's sponsoring surfboard-making industry). *Work space and support facilities for small business:* This aims to establish a privately or publicly owned project using a work house building or the like to provide work space and a support framework for small firms with common facilities (e.g., reception, telephones, telex, cleaning, maintenance, conference rooms).
Small business and new venture start-ups	*Community business advice project:* This is an organization formed to provide professional advice and assistance to small businesses in the areas of management, marketing, accounting, financing, locating resources, and problem solving. One such organization is a nonprofit organization run by a committee that decides which clients to help and how much to charge them (e.g., no charge for low-income groups and marginal businesses and standard consultant fee to established firms). It employees a full-time business adviser who works directly with clients in their home or place of business.
Business expansion/ retention	*Export promotion scheme:* This is a scheme to assist small- and medium-sized firms in exploring new export markets by representing them at trade fairs and missions. *Regional technology center:* This is a center to research and identify technologies appropriate to the region's socioeconomic base and to help people who are developing these techniques start up a business. It provides the necessary backup support facilities and management advice.
Local government	*Joint public-private ventures:* Joint venture agreements for construction of business venture facilities to revitalize the central business district generate employment and create revenue for the city government. These facilities could include a cinema complex or an administrative center with shops, parking garage, and sports facilities.

to attract and retain firms as a major component of its economic development strategy.

References and Suggested Reading

Barnes, J., and A. C. Noonam. 1982. Marketing Research: Some Basics for Small Business. *Journal of Small Business Management,* July.

Bennett, E. 1987. *Social Intervention Theory and Practice.* Lewiston, NY: Mellen.

Erickson, Rodney. 1987. Business Climate Studies: A Critical Evaluation. *Economic Development Quarterly* 1 (1): 62-71.

Finsterbusch, L., and C. P. Wolf, eds. 1977. *Methodology of Social Impact Assessment.* Washington, DC: Environmental Design Association.

Frieden, Bernard. 1990. City Center Transformed. *Journal of the American Planning Association* 56(4).

Garner, Linda. 1983. *Community Economic Development Strategies: Creating Successful Business: Vol. 2. Choosing the Business Opportunity.* Berkeley, CA: National Economic Development and Law Center.

Goodman, M., and E. Love. 1982. *Project Planning and Management: An Integrated Approach.* New York: Pergamon.

Green, Roy E. 1991. *Enterprise Zones: New Directions for Economic Development.* Newbury Park, CA: Sage.

Harty, H., et al. 1977. How Effective Are Community Services? *Procedures for Monitoring the Effectiveness of Municipal Services.* Washington, DC: Urban Institute.

Harty, H., et al. 1981. *Practical Program Evaluation for State and Local Governments.* Washington DC: Urban Institute.

Hines, M. 1976. *Principles and Practices of Real Estate.* Homewood, IL: Irwin.

LaMore, R. 1986. *An Introduction to Financing Community-Based Economic Development.* East Lansing, MI: Center for Urban Affairs.

Lewis, Sylvia. 1993. The Bank with a Heart. *Planning Magazine* 59(4).

Litvak, L., and B. Daniels. 1979. *Innovation in Development Finance.* Washington, DC: Council of State Planning Agencies.

Lockhart, A. 1987. Community Based Economic Development and Conventional Economics in the Canadian North. In *Social Intervention Theory and Practice,* edited by E. Bennett. Lewiston, NY: Mellen.

Mahood, S., and A. Ghosh. 1979. *Handbook for Community Economic Development.* Sponsored by the U.S. Department of Commerce, Economic Development Administration, Washington, DC. Los Angeles: Community Research Group of the East Los Angeles Community Union.

Mancuso, J. 1983. *How to Prepare and Present a Business Plan.* New York: Prentice Hall.

McCarthy, J. 1981. *Basic Marketing: A Managerial Approach.* Homewood, IL: Irwin.

Parzen, Julia Ann, and Michael Kieschnick. 1992. *Credit Where It's Due: Development Banking for Communities* Philadelphia, PA: Temple University Press.

Planning Exchange and Project North East. 1985. *Group Marketing and Export Assistance for Small Businesses.* Glasgow, UK: The Planning Exchange.

Richard, J. W. 1983. *Fundamentals of Development Finance: A Practitioner's Guide.* New York: Praeger.

Rubin, Barry, and Margaret Wilder. 1989. Urban Enterprise Zones: Employment Impacts and Fiscal Incentives. *Journal of the American Planning Association* 55(4).

Seley, S. E. 1984. Targeting Economic Development: An Examination of the Needs of Small Business. *Economic Geography* 57(2).

Stevens, R., and P. Sherwood. 1982. *How to Prepare a Feasibility Study.* Englewood Cliffs, NJ: Prentice Hall.

Stokes, Barry. 1985. Determination Helped Snatch this Town from the Jaws of Economic Death. *National Journal,* December 21.

Western Community Action Training. 1970. *Economic Development Manual.* San Francisco, CA: Social Change Press.

9 | Human Resource Development

Human resources are pivotal to the economic development process. But although they are essential, certain individuals and groups in society are not able to compete in the emerging growth sectors of biotechnology, electronics, information, computing, and health industries. The reasons for this have more to do with access to and expectations of these sectors than with the desire of individuals to work.

In some instances, jobs are literally moving away from people. That is, the new locations for jobs are physically far away from where people need work. In other cases, jobs are transforming or disappearing faster than the population can adapt. The clearest evidence of this is the rapid decline in the labor input to the manufacturing sector. In the food/retail sectors, many former jobs such as check-out clerk are being reduced or abolished as a result of the introduction of new automation machinery.

Rhetoric about human resources being pivotal to economic recovery in the United States is far better than our performance. For example, Figure 9.1 illustrates the difference in American versus Japanese investment in autoworkers' training. As this figure indicates, the quality of Japanese vehicles is not just in better machinery.

The few data available indicate that American firms invest between $30 and $44 billion in training and staff development. This is only 1.2% to 1.8% of private sector payroll. As a result, the nation is not preparing its human resources for competition today or tomorrow.

As jobs are being shed, people remain an important community resource and are crucial to a locality's economic recovery. The United States has tended to place more of its national resources in income-support programs than in job-creation and maintenance schemes. In contrast, other developed nations, such as Sweden, have recognized the need to make job creation the key component of their labor market programs in times of contracted economic activities. This need is only now being recognized here. Despite improved national employment figures and recent employment-generation programs initiated with the Job Training Partnership Act and its predecessor CETA, the United States has not made job creation a national priority other than through macro-economic measures. The Clinton administration is seeking to remedy this situation by implementing a series of new measures that make work and the creation of work opportunities the central thrust of government policy. As a result, new opportunities for state and local government are being designed to enhance collaboration with the federal government in the arena of human resource development.

Matching Human Resource
Development Programs and Objectives

A number of initiatives can improve employment opportunities. Here I concentrate on people-based initiatives that directly improve the individual's access to jobs or skill development. Human resource development, Michael Swack suggests:

> recognizes that one of the most valuable resources in a community is its people. Community economic development . . . provides them with the tools and knowledge to control the resources of their community . . . as a function of providing technical skills and knowledge to people already working in a community so they can carry on the effort of development. (Swack and Mason 1987, 343)

Human resource development initiatives can be divided into four broad categories: vocational training and education, job placement projects, client-oriented job creation, and job maintenance activities. Table 9.1 displays the various human resource development tools in these categories, which are discussed below.

Other employment development initiatives that have a human resource orientation include but are not limited to job sharing, job pairing and part-time employment, and a move to phased-in retirement along with flexible, shorter hours of work and new two-tiered wage scales with lower starting wages. These initiatives are not discussed here, however, because local government is not in a

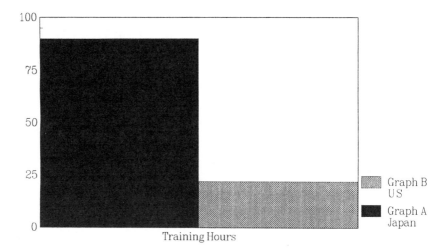

Figure 9.1. Annual Hours Training per Autoworker
SOURCE: Data from Krafick, 1990.

position to initiate shifts in work patterns. This is a subject for discussion and implementation by other tiers of government, by unions, and by employers.

Customized Training

The concept of customized training is relatively simple. It provides training either at the employer's site or at the local college specifically designed to meet employer expectations. Human resources can thus be used to attract business because of reduced staff development costs.

Although this type of program is usually reserved for new firms moving into an area, it may also be used by existing firms that are expanding. One of the most innovative customized training programs in the nation is the Ohio High Unemployment Population Program (HUPP). This program targets training and employment services to chronically unemployed minority males. In this program, the employer is used as a home base or site for a combination of on-the-job training and education for a credential or degree for each participant. The basic idea is for the participant to develop job habits and a job resume while gaining a degree. Although it is too early to evaluate this approach, early returns indicate that it is working and attracting hard-to-reach long-term unemployed individuals.

First-Source Agreements

First-source agreement refers to a contractual obligation entered into by firms that receive substantial government assistance such as tax holidays and government loans. Under the contract, the assisted firm agrees to interview, as the "first source," local people for the available positions. After reviewing them, the employer may select anyone for the vacancy. This agreement generally stipulates the referral source and the number of referrals required before the employer can place advertisements or interview persons from outside the specified geographic area.

In some instances, firms will request local referral voluntarily as a civic obligation in high unemployment areas. These voluntary efforts are also used to stimulate employers to look at local youth as a first source. Youth-oriented first-source agreements generally have a "job try-out" component.

Employment Maintenance

Employment maintenance or support programs use unemployment funds to develop jobs for hard-to-place people. In this scheme, a local government or nonprofit organization receives the welfare payment as a subsidy for employing a person at the full wage level. A large number of community development corporations have employed this approach with considerable success. This type of project is one of the keystone initiatives of the Clinton administration. Many urban and rural local governments have indicated an interest in this type of program as a means of restoring teachers aides, playground supervisors, assistant librarians, and many other para-professional positions.

Skill Banks

This is a relatively simple concept, aided by the advent of the microcomputer. The idea is that the employment officer, or a local volunteer group or other person(s), collects refined information on the skills and backgrounds of local unemployed persons. This

Table 9.1   Human Resource Development Tools and Techniques

Tool	Goal			
	Vocational Training and Education	*Job Placement*	*Client-Oriented Job Creation*	*Job Maintenance*
Customized training	X		X	X
First-source agreements		X		
Supported work program				X
Local employment officer	X	X	X	
Skill banks	X	X	X	
Training programs	X	X		
Youth enterprise	X		X	
Self-employment initiatives	X		X	
Disabled skills development	X	X	X	

skill inventory can be used for a number of purposes. First, it serves as a source of community data on the nature of the job development needs of the unemployed. Frequently, communities know that there is high unemployment, but they do not know what kinds of skills the unemployed have or the kind of jobs they need. The skill bank helps address this problem. Second, the dominant use of the skill bank is its inventory of skills that can be matched to available vacancies in the local area. Finally, skill banks collect unemployed people into cooperatives or community-based employment initiatives. A skill bank is an invaluable tool for assessing local abilities and building new jobs or services on the basis of these skills.

In order to initiate a skill bank, organizers should first do a survey to document the demand for various occupations or skills and the availability of skilled labor to fill those needs. As a second step, the existing training and education facilities in the region should be assessed by their accessibility, suitability, and ability to meet further labor market needs. Recommendations for upgrading, extending, or modifying education/training programs have been made on the basis of these studies.

Training Programs

There are a plethora of training programs across the country, organized in a variety of ways in different states. Most contain provisions for special courses to meet the needs of local employers and include upgrade training, apprenticeship courses, retraining of dislocated workers or of women reentering the workforce, as well as traditional course offerings. Local training councils should be formed to integrate these efforts with local development strategies. For example, communities that embark on tourism strategies should develop courses that provide initial education and upgrade the skills of those already in the hospitality industry. Generally, local employment and training boards known as Private Industry Councils (PICs) will be more than happy to cooperate with such projects if they are aware of the economic development goals of the community. Wolman, Lichtman, and Barnes (1991), after examining education, training, and skills, conclude that training in a specific skill area does have a significant payoff for the underskilled.

Private Industry Councils

There has been a continuing gulf between employment and training efforts and economic development, particularly for the disadvantaged. In fact, training has been a "revolving door" for many people. They attend training courses that frequently lead to dead-end or unsatisfying occupations, only to be sent to additional training as a remedy for their unemployment. Recently, community groups have attempted to merge the employment and training process with local job creation. The device for accomplishing this are the PICs. These boards pull together local business and community-based economic development programs into a single employment network.

PICs attempt to coordinate the frequently fragmented efforts of parties interested in employment within a region. Although they vary in composition, they generally include representation from key local firms, local government, state government, unions, and educational institutions. The members of the board must commit themselves to creating more flexibility in both training systems and job entry arrangements.

They must pull together all the government-sponsored training initiatives and integrate them with local economic development programs. One of their major responsibilities is to identify and encourage new ventures that have skill requirements readily met by local people. Another is to match local training activities with the needs of local employers.

The types of activities undertaken by employment and training boards depend on the local situation. Generally, they involve themselves in these aspects:

- Managing workforce training schemes such as group apprentice schemes, adult retraining, and new training initiatives
- Offering business support services that increase employment
- Developing shared facilities for training activities
- Operating general literacy and community education projects
- Developing on-the-job training and work experience activities for young adults
- Attempting to reduce discrimination in employment for disadvantaged groups

Community employment programs (which are discussed in the previous chapter) offer an appropriate mechanism for coordinating economic development and employment. They offer a vehicle for communities that wish to combine the community-based concept with more traditional economic development activities.

Youth Enterprise

There is growing interest in helping young people to become acquainted with business development. In part, this is an attempt to rekindle entrepreneurial spirit in the nation's youth. It also represents a new emphasis on helping people create jobs for themselves rather than waiting for jobs to be created for them.

Youth enterprises have been started through the Community Youth Employment programs as well as various service clubs and community-based organizations. These enterprises vary in size and focus, and a few have become major self-supporting enterprises.

The basic idea is to start businesses within the skill range of youth to give them some experience in how the business system works. Community-based groups have considerable resources to foster such activities. They can offer vacant space or equipment to the youth enterprise as well as mobilize service clubs and others to provide the technical assistance for the projects.

Self-Employment Initiatives

These initiatives attempt to assist unemployed people of any age to create employment ventures themselves using their own labor as the primary resource. They include small business start-up efforts in which the unemployed receive a small loan and/or grant to initiate a new, independent, community-serving business. During the period that the business is being organized, the unemployed person receives an allowance equivalent to the unemployment benefit. This allows the enterprise to reach the stability point before wages have to be paid.

There also are programs available that support self-employment ventures organized by a group rather than by an individual. In these projects, such as the micro-loan programs discussed in the previous chapter, a specialist works with groups to examine opportunities and exploit them using unemployed persons as the basic resource.

Disabled Skills Development

Physically, emotionally, and educationally disabled persons have always had difficult placement problems. Intervention in these situations ranges from sheltered workshops to employing a special disability placement officer (DPO).

The DPO is responsible for seeking out employment opportunities for disabled persons that fit their capabilities. With modern equipment, there are more of these opportunities than most employers realize. The DPO's job then is to see how the work environment can be made to fit the person rather than the person fitting the job. The DPO also provides employers with information on assistance schemes for disabled persons and their employers. Local governments can assist DPOs and the disabled by being positive examples, such as including the disabled in the development planning process and by allowing council offices to be used by the DPO or others working on issues of the disabled.

Illustration 9.1. The Machine Action Project

The Machine Action Project (MAP), located in Springfield, Massachusetts, exemplifies how labor's involvement can lead to innovative local approaches to training and retraining and manufacturing retention. MAP is an affiliate organization of the Federation for Industrial Retention and Renewal (FIRR), a network of labor-community coalitions committed to maintaining and developing the U.S. manufacturing sector and the high-paying jobs it provides.

The FIRR model of industrial retention focuses on three issues: future ownership, integration of minorities and women, and maintaining manufacturing competitiveness. FIRR recognizes that if the U.S. manufacturing sector is to become more competitive, new high-performance forms of workplace organization will have to be adopted, which will require a substantial investment in worker training.

MAP's involvement in training emerged in response to restructuring of the local metalworking industry. Without MAP's efforts at identifying and developing a response to a skills mismatch in metalworking firms, this essential part of the Springfield area economy would have been lost. The effectiveness and uniqueness of their approach has been recognized widely. In 1988, MAP was the recipient of a Ford Foundation-Kennedy School of Government award as one of the ten most innovative community projects in the country. In 1990, MAP received another innovation award from the Arthur D. Little Foundation.

The late 1970s and early 1980s were a time of intensive restructuring of the metalworking industry in the Springfield, Massachusetts, area. Large, highly automated firms were closing and small producers employing flexible production techniques were growing. Approximately 15,000 metalworking jobs were lost between 1980 and 1985, virtually all in large facilities.

In 1985, the local International Union of Electrical Workers (IUE) union members at the American Bosch plant in Springfield began to notice signs of disinvestment. Through business manager Bob Forrant, the union began warning state and local political representatives and the local economic development community that this plant employing 1,000 workers was likely to close. The response from all quarters was that of disbelief. The IUE was told their early warning indicators were responding to a false alarm.

When American Bosch announced its closing in 1986, dismantling and movement of production offshore was already in process. Although union activists were able to negotiate a better severance package, they were unable to convince local economic development actors to support an eminent domain takeover of the facility and were frustrated by their lack of influence. The union then worked with the local private industry and labor councils to seek state funding for designation as a State Industry Action Project. Funds were awarded in 1986 to organize MAP to conduct local labor market research and to develop worker training strategies. MAP's research revealed a previously unidentified skills mismatch. While much attention had been paid to the loss of major metalworking employers, many small producers were thriving. In contrast to the large automated facilities, the smaller shops employed flexible production processes. Workers in small facilities needed higher skill levels, requiring them to set up and operate three or more machine tools, to read and assemble parts from blueprints, and to maintain their own tools. Few workers in large automated plants possessed these skills. A key difference between these and the central government programs is their focus on retention and expansion of manufacturing in order to maintain economic competitiveness and high-paying, blue-collar jobs.

Thus, workers displaced from the large firms did not have the skills needed by the small flexible producers. The lack of skilled labor to feed the small producers was also a result of a general perception in the area that metalworking was a dying industry, and that blue-collar work was inferior to service occupations. As a result, new labor force entrants did not consider employment in metalworking.

In 1988, MAP was funded by the Massachusetts Department of Education to offer adult education programs in five vocational high schools in the region. The courses developed were targeted to upgrade the skills of those employed in area metalworking shops, although they also served the unemployed and high school students as well. MAP entered into a memorandum of agreement with Springfield Technical Community College, the local chapter of the National Tooling and Machining Association, and several federal, state, and local agencies on a common curriculum to be used by all local metalworking training institutions. MAP assumed a coordination and informational role in the agreement, conducting research on the region's training needs, developing recruiting strategies to target women and racial and linguistic minorities, securing funding, and following up on trainees.

The program served the needs of the small metalworking shops, the employees, and the schools. The firms paid minimal fees for state-of-the-art instruction they could not provide on their own. The coordination of courses, a common intake assessment, and a competency-based curriculum allowed trainees to take courses with different providers without losing program continuity. Further, the articulation agreement between the schools allowed trainees to accumulate credits toward an associate's degree in metalworking technology. The state funds and tuition were used to update the schools' curricula with one recently approved by the National Machining Association and to buy equipment. Further, MAP co-sponsored a summer institute on new manufacturing technologies for 20 local secondary vocational education teachers to keep them updated in the latest manufacturing technologies.

MAP also received U.S. Department of Labor Women's Bureau funds to develop a program to increase the presence of women in the machine trades. MAP commissioned the Massachusetts Career Development Institute (MCDI), a nonprofit training provider targeting low-income youth and adults, to provide the training. MCDI already offered its trainees a number of support services, including child care, to eliminate barriers to program completion.

The Inverclyde Training Trust

More than any other community on the River Clyde, Inverclyde (consisting largely of the towns of Greenock and Port Glasgow) depended most heavily on shipbuilding. Even after a protracted period of shipbuilding decline, shipbuilding still accounted for 17% (roughly 6,500 jobs) of all employment in Inverclyde in 1981. By the end of the 1980s fewer than 1,000 shipbuilding jobs were left. This is a dramatic decline bearing in mind that the area's major shipbuilder, Scott Lithgrow, alone accounted for 9,000 workers in the mid-1970s.

The impact has been highly localized as there was a very strong workplace-residence overlap due to the historical tradition of shipbuilding in the area, and its relative isolation from the larger Glasgow area. The unemployment situation in Inverclyde deteriorated dramatically over the period of this job loss. In the early 1980s, the unemployment rate was roughly 20% above the Scottish average, and by the late 1980s, it was 50% higher. This was despite a population decline from 105,000 to 98,000 between 1976 and 1985.

A number of responses were developed in an attempt to combat the decline in shipbuilding employment. A key innovative measure was the development of Inverclyde Training Trust (ITT) in 1986. Although control of ITT was local, the initial funding for its activities was provided by a regional economic development organization, the Scottish Development Agency. ITT organizers realized that the jobs lost in the shipbuilding industry would not be recovered, due to international overcapacity. Further, they recognized that in a stagnant national and regional economy, possibilities for attracting mobile capital were limited and competition between areas for that capital was fierce. Thus their emphasis was on providing displaced workers with new skills to equip them for alternative employment or self-employment.

To link training and local economic development, ITT focused on providing skills for the unemployed both to fill existing vacancies and to attract inward investors, and for the employed to enhance the productivity and performance of local companies. To accomplish its goals, the organization would have to convince local unemployed people and employers of the benefits of training and retraining, and obtain the resources to make local training programs more effective.

Two features of ITT enable it to perform a difficult and innovative role in the local economy. First, it was established as a training broker, and not as a provider of training. ITT sits in the middle of Inverclyde's training system, establishing the needs of the local unemployed and employers and trying to find means of meeting those needs. In a sense, it is a catalytic organization whose presence in the local system generates dynamic change or development in training. Second, it is a freestanding organization with only minor core funding and resource support. ITT essentially trades in training services. In this capacity it makes use of existing U.K. national and local government programs as well as training programs funded by the Social Fund of the European Common Market.

A major part of ITT's training portfolio is supported by the Social Fund. Many trainees are eligible for a training allowance, and a survey of trainees revealed that 25% would not have been able to participate without this support. This distinguishes ITT's training from the "benefit plus" regime of the national program, ET.

ITT's 95% completion rate indicated that the thorough monitoring and feedback of participants is a critical component of the program. The post-training placement rates, however, reveal a less positive outcome. Whereas half of the former shipyard workers were either employed or in further education or training, the corresponding figure for trainees from other industrial backgrounds was only one-third. The difficulties in shifting people from unemployment to self-employment are evident, although the figures are somewhat better than those achieved in some of the large-scale redundancies elsewhere in the U.K.

The high rates of unemployment vividly illustrate how training itself is not a solution to the problem of redundancy or unemployment. It can facilitate the redeployment of reasonable numbers and help the local unemployed generally to find work, but it needs to be accompanied by corresponding measures to replace some proportion of the net job loss resulting from the redundancy. This conclusion is supported further by a survey of trainees that indicated that only 50% of former shipbuilders felt that their training helped them directly in getting a job. Nevertheless, 91% valued the ESF training sufficiently to recommend it to other unemployed people. This figure indicates that workers obtain value other than direct and immediate employability from their training. Indeed, although nearly 60% of former shipyard workers saw the acquisition of job-related skills as the key advantage of training, 33% stressed enhanced personal confidence. For many trainees, personal confidence and related factors significantly outweighed job-related skills as the key advantage. By improving morale and self-esteem, training keeps the local workforce in a position to take advantage of new jobs that may locate in the area. As Inverclyde is hoping to attract additional employment as a result of the establishment of the Inverclyde Enterprise Zone, maintaining motivation is critical.

SOURCE: Fitzgerald and McGregor 1993. Reprinted with permission.

Case Study 9.1. Employment Planning as Economic Development

How does an older Northeastern city with a declining industrial base, high unemployment, uncoordinated federal economic programs, and a reputation for poor service in job training and placement overcome its handicaps and use its federal resources to build a job training and referral system that draws industry back into town, creates jobs, and wins the respect of private industry?

In Eastport, which has come a fair way toward achieving these goals, the answer rests in a combination of the city's commitment to provide real job service to the private sector and its determination to save neighborhoods by seeing that there are jobs for the people who live in them. Indeed, the key to the economic turnaround of Eastport is jobs, especially new jobs in the heart of the city to replace those lost. The blueprint for achieving this is Eastport's Employment Development Plan, with a realistic approach to urban planning and a coordinated use of federal funds. The plan's principal goal is creating jobs and assuring people's access to them.

The essence of the Eastport Plan is the coordination of the efforts of all the federal urban economic stimulus programs into one local agency used to build a base for both short- and long-term improvements in the city's economy. Federal funding for public works, economic development, and training and employment had been coming into the city under separate programs. The plan provides a means to combine these programs at the local level. Its goal is to create 14,000 new permanent jobs over 5 years.

The city of Eastport had some successful experience in downtown revitalization to draw on in putting together the plan. The city had already faced all the problems of many older cities—a declining inner-city taxbase, movement of the middle class to the suburbs, increased demands for social services by those remaining, and a growing unemployment problem. As one response to its problems, the city chose to make a direct investment in the downtown waterfront, transforming this decaying and obsolete commercial area into a housing, commercial, and tourist center that has attracted national interest. The success of the downtown revitalization effort demonstrated to residents that the city could be made livable again. The city proposed to revitalize several neighborhoods. Residents, who had seen the downtown successfully restored, had reason to hope that the new plan might work in their neighborhoods.

One of the first steps the city took was to reorganize the city's various employment and economic development activities into a single agency, the Employment and Economic Policy Administration (EEPA), and to give it responsibility for providing new or expanding industries in the area with trained workers or federal job training program enrollees whom a firm could train.

Before the city could undertake job training and referral services for the private sector, it had a major handicap to overcome—its reputation for providing poor-quality employment services to those industries. Private employers felt that the city had provided inadequate job-orientation services to the potential workers it referred. Many trainees dropped out of the program in which they were placed, or the employers were so dissatisfied with them that they did not go back to the program for more. There were few links between the city employment program and the community organizations that understood the special needs of the cultural and ethnic groups living in the inner city, from which many of the prospective workers were drawn. Private employers felt that the city did not give a high priority to the services it provided.

The EEPA set out to correct this situation. The city sponsored a survey of employers to find out what specific problems they had experienced in the past, and asked for their advice on how to improve the situation. One problem frequently identified was a lack of direct involvement on the part of the city officials in dealing with private-sector industry. To correct this, the city first transferred on-the-job training activities from the independent agency that had been acting for the city to the EEPA. The EEPA then formed a coalition with the National Association of Business (NAB) and the Urban League to provide the close contacts among the city, jobs, and people, which were lacking before.

Later, the on-the-job training program of the EEPA formally merged with the NAB. The two agencies are now housed together, each agency maintaining its own role in the employment process. The NAB staff aggressively approached the business community to find jobs for the economically disadvantaged. Firms that agreed to participate could hire individuals directly or through an on-the-job training program. The payoff to the firms is the screening, job orientation, and employability counseling provided to each person by the city agency before the initial employment.

One illustrative program is a training program under way to provide long-range training opportunities in the marine and computer fields. The harbor is the site of a major new industrial park that will provide jobs for 3,500 skilled and unskilled workers. Using a variety of funding sources, the city renovated and equipped one of the buildings in the park to serve as a job-training center—the only vocational training facility directly linking the growing marine industry with a skills training program.

The center's staff identified specific highly skilled occupations that required no more than 10 months of training. They then verified the need for workers with these skills with the firms operating the Marine Industrial Park. Under the agreements entered into between a potential employer and the center, each employer must appoint a training supervisor to assist in the design of the training programs. Each of the participating firms must use the plan developed as the basis of the agreement with the job training center. The job trainees are trained either at the center or on the job site. All employers using the center's services must make a commitment to hire first the individuals trained through the center program. This has not been a problem, because the agreements provide for heavy employer involvement in the specification and design of the training courses in all skill trades with known labor shortages.

Eastport seems to be well on its way to a healthier job base than it has had in some time. But there is more to the plan than the link between jobs and economic development, although that is a key element. Through the strengthening of the city's economic base and by providing jobs to unemployed city residents, the city's leaders believe they will reduce the problems of crime, poor housing, inadequate health facilities, and inaccessible recreational facilities.

One of the most exciting features of the plan is the one providing unemployed workers with loans to start their own businesses. Under this program, the unemployed continue to receive their unemployment insurance payments for a full year with a supplemental payment while they start their new business. Participants must take up to 6 weeks of small business training before they open a business and become involved in the SCORE (Service Corps of Retired Executives) program for technical assistance to their new firm. The program has received a lot of criticism from local merchants who claim that their firms were not subsidized at the outset and that these new businesses represent unfair competition. However, the program guidelines are relatively strict on the new firm competition area by restricting new start-up firms to business types not represented in the local community. What do you think of the complaints of the local businesspersons and how might their issues be resolved, if at all?

Case Study 9.2. Moving the Poor Off Welfare

Many cities are experiencing very different economic circumstances as a result of the national economic transformation. While the Midwest has a high unemployment rate and large concentrations of poverty, both the Northeast and the South are booming economies with relatively few jobless people. All of the cities agree that they must initiate programs to move substantial numbers of chronically poor people off welfare and into relatively well-paying jobs.

The secret to achieving this, some experts suggest, is to involve the poor in counseling, education, and training programs far more intensive than any attempted in the past and to guarantee continued support services, such as medical insurance and child care, until they are secure in their jobs.

Margarita Majias was one of 30,000 welfare clients who officials say found employment under the state's program in the past 3 years in

Holyport, a California city still suffering from the closing of industrial plants. As a single mother with one child, she had tried working, with a job in a shoe factory for about $4 an hour. Soon, she said in an interview, she found she could do better by going on welfare, which also provided food stamps and medical insurance through Medicaid.

After several years on welfare, she was persuaded by state counselors to enter training, which included instruction in how and where to apply for work. Eventually, she found a job as assistant property manager for an apartment complex. Her income went from $500 a month on welfare to $960, along with a free apartment in the complex and a promise of raises in the future, with benefits.

"It's good work, and gives me a sense of purpose because other people are depending on me. Sure it's hard, but I like it," she said. "And I don't have to worry about medical insurance for my daughter. It is provided with respectability for the first time in my life."

Few experts believe these efforts will be successful with every welfare case. But most agree that long-term welfare dependence can be reduced substantially by innovative efforts, which may eventually help to break cycles of poverty, broken families, crime, drug addiction, and teenage pregnancies.

The Holyport program is voluntary, unlike some other cities' programs. Eligible applicants can elect regular welfare payments or a stipend for the work performed in lieu of welfare. Most choose the check because it is a symbol of moving away from welfare. The program includes close counseling, education, training if needed, and job placement with the promise that Medicaid, the medical insurance program for the poor, will continue for a year if the employer does not provide it. That is important because many women say they would rather stay on welfare and receive Medicaid than take a job paying marginally more that does not have medical insurance.

In addition, Holyport has added a new wrinkle to the current efforts. A proposed program called Work Assistance allows a select group of nonprofit Community Development Corporations (CDCs) to hire and train welfare recipients at subsidized wages. The Holyport Spanish Speaking CDC operates a gasoline station, a community supermarket, a rental car agency, and several health clinics, as well as a day-care center. Under the plan, the city will subsidize the employment of welfare recipients in all firms operated by the Holyport Spanish Speaking CDC. The CDC must pay above the minimum wage and guarantee to place the participants in unsubsidized jobs within 6 months.

At the last meeting of the Holyport Employment Council, Ernesto Gutterez asked that the subsidy be raised and the length of time increased because the quality of applicants was so low that more remediation was necessary. The Employment Council is upset by the request because it feels that the program is really too expensive and that the Spanish Speaking CDC is using too much of its funding for English and literacy instruction. Based on an assessment of the local, primarily Hispanic welfare clientele, would you consider the Spanish Speaking CDC's request reasonable? If so, why? You might look at some local data and interview welfare and community organizations to determine the job readiness of the local welfare population. What would it take to put most of your local limited English-speaking welfare population into unsubsidized employment?

Case Study 9.3. Young Need a Job—Make Your Own!

Tung Ng was offered a job as a Vietnamese youth director by the Viet Family Life Center but wasn't sure if the ideas he had for the job would be acceptable to the organizers of the Viet Family Life Committee. That was 2 years ago and Tung now draws $2,000 a month for work that has become a mission.

Tung's idea for the largely Vietnamese and Cambodian, lower-income Crestview Park neighborhood was a program that would teach youngsters 14 to 21 about business, personal money management, and entrepreneurial savvy. Tung brought a lot of business know-how with him from Viet Nam. He had worked for the U.S. Air Force in the base store. His family ran restaurants and variety stores in Saigon. He felt that the American-born and -raised Vietnamese and Cambodian youth were using their entrepreneurial skills to form gangs instead of starting their own businesses.

Tung started his program, called Youth Action, with only 20 youngsters. Based at a local center, the program is open to any child in the community. Only about 30% of its members belong to the Viet Family Life Center. Tung admits Youth Action has many components similar to Junior Achievement. However, Youth Action is more professional because it is geared to help young people actually own and operate businesses in their communities. It differs too from standard job training programs because it doesn't teach them how to do a particular task in someone else's business. Youth Action shows youths how to develop "minibusinesses" of their own. They learn by attending a 10-week training program after school and on weekends.They also act as interns in jobs they can accommodate year-round, after school, and on weekends.

At the conclusion of the training program, each youth has a business plan. These plans are screened by a panel of local Asian businesspersons. Those that meet the test of business worthiness are seed-funded with a grant of $2,000 and a small loan. The business must be in community-related areas and fit a community need. The youth cannot merely find a nice white-collar occupation outside the community as the answer to his or her problem.

"We live in a world where even many white-collar jobs in big and smaller companies are becoming scarce for experienced, educated adults," Tung points out. "Asian youngsters must be trained early to turn their talents and skills into jobs they create for themselves."

The basis for Youth Action, says Tung, is "working to re-create historic values of building something for yourself and fostering the self-dignity, independence, and pride essential to long-term achievement." Tung hopes to expand the program if it continues to be successful.

The Family Life Center provides facilities for Youth Action activities but takes no money from them. All the money Youth Action generates is used to support its projects and for the youngsters' salaries. Members must give 15% of their earnings back to Youth Action. These funds are held in a trust account for each member's future education at the college or vocational level. A person who does not go to college forfeits his or her accumulation.

Since the program began, the teenagers have launched a prospering catering business, a computer training center, a word processing service, a photography business, a summer tutorial program for elementary schoolchildren, and an answering service. In addition to those ongoing services, Youth Action has sponsored several one-time projects such as yard sales, breakfasts, bake sales, and an ice cream store day (they made the ice cream at a local outlet and sold it). In the planning stages are a baby-sitting service, a Vietnamese and Cambodian book fair to encourage cultural restoration among inner-city youngsters, and a children's computer and electronic swap meet. To encourage sound money management, members must plan a personal budget and establish a savings account at the Korean-owned Community Savings & Loan Association. That bank was chosen, Tung noted, because he wants the youngsters to get in the habit of supporting local Asian business.

"The bank officers know them so well now that they've gone a long way toward establishing a sound relationship with a bank," said Tung. "I explained to them how important that will be when they're [old enough] to apply for credit and need good references."

"I've found that most of these kids have lots of talent," Tung said. "They all have things they can do. If I say I need a sign made, or whatever, someone will say 'I can do it,' and they do it."

The adult board of Youth Action wants to consolidate all businesses under a single corporate or limited partnership status. The rationale for this move is that the business license and other procedures are too complicated and make business start-up more difficult than business operations. Several boardmembers counter that this is the essence of the experience. Tung objects to this change. The cost of opening these ventures is time-consuming and creates a liability for the Family Life Board. Assume the board has come to you as an economic development expert and asked you to chart an alternative plan for the youth businesses that is less expensive and reaches more youth.

Table 9.2   Human Resource Development Project Idea Starters

Tool	Project Idea Starter
Local employment officer	*Self-employment course:* A project that gives advice and encouragement to people wishing to create their own jobs by setting up a small business. Activities could include running workshops in self-employment such as on the following: • Evaluating business ideas • Bookkeeping for small businesses • Financial budgeting in small businesses *Helpmates:* A project that aims to teach mildly intellectually disabled young adults the necessary skills to work in the area of gardening services and providing various services to people who are aged, physically handicapped, or infirm.
Unemployed/ disadvantaged venture creation	*Horticultural service for businesses:* Unemployed young people do interior landscaping for local businesses and lease to these businesses plants that the service has grown and will maintain. *Women's affirmative job action project:* A project designed to improve women's access to jobs by: • Analyzing the local economy and identifying employment opportunities and the skills required by these opportunities • Compiling database on skills possessed by women in the community • Matching skills required by employers to skills possessed by the women • Formulating suitable projects for upgrading and diversifying skills as necessary • Reviewing employment practices • Developing programs to remedy discriminatory practices and to encourage women to apply for a wider range of jobs • Promoting job sharing and other flexible work agreements among potential employers of women *Craft facility:* Space and facilities (e.g., kilns and looms) provided for craftspersons to enable them to develop their crafts to market standards and volumes and to sell from that space.

Summary and Conclusion

Unemployed persons are generally viewed as incapable. This is far from the truth. The unemployed are a resource. They can help themselves as well as the community. (See Table 9.2 for a sampling of human resource development projects.) The programs discussed in this chapter are usually auxiliary or adjunct to other efforts. That does not mean that they are unimportant or that they should be an afterthought. They can, especially in the short term, be a major component in the overall local/ regional economic development strategy. Irrespective of whether or not a council is part of a regional scheme, there are opportunities in this area that can be taken to meet the particular training and employment needs of community members.

References and Suggested Reading

Ashenfelter, O. 1978. Estimating the Effect of Training Programs on Earnings. *Review of Economics and Statistics* 60 (1): 47-57.

Becker, G. 1975. *Human Capital,* 2nd ed. New York: Columbia University Press.

Blakely, E. J. 1982. Economic Development and Job Creation: Some Ideas and Examples for the United States. In *Job Creation Through the Public Sector? A Strategy for Employment Growth,* edited by Peter Hollingworth. Melbourne, Australia: Brotherhood of St. Laurence.

Blichfeldt, G. 1975. The Relations Between School and the Place of Work. In *School and Community,* edited by G. Blichfeldt. Paris: CERI/OECD.

Choate, P. 1982. *Retooling the American Workforce: Towards a National Strategy.* Washington, DC: Northeast-Midwest Institute.

Field, F. 1977. *Education and the Urban Crisis.* London: Routledge & Kegan Paul.

Fitzgerald, Joan Race. 1992. *Job Training and Economic Development: Barriers to Equity Planning.* Unpublished manuscript, University of Illinois, Chicago.

Fitzgerald, Joan Race, and Allan McGregor. 1993. Labor-Community Initiatives in Worker Training. *Economic Development Quarterly* 7 (2): 176-179.

Furst, L. 1979. Work: An Educational Alternative to Schooling. *Urban Review* 11(3).

Gleaser, E. 1980. *The Community College in the United States.* Paper presented at an OECD Conference, Higher Education and the Community, Paris, February.

Goldstein, G. 1977. *Training and Education by Industry.* Washington, DC: National Institute for Work and Learning.

Goodman, R. 1973. *After the Planners.* New York: Simon & Schuster.

Hamilton, M. 1980. On Creating Work Experience Programs: Design and Implementation. In M. Hamilton (Ed.), *Youthwork National Policy Study.* Occasional Paper No. 3. Ithaca, NY: Cornell University.

Holloway, W. 1980. Youth Participation: A Strategy to Increase the Role of In School Youth in Creating Job Opportunities. In W. Holloway (Ed.), *Youthwork National Policy Study.* Occasional Paper No. 3. Ithaca, NY: Cornell University.

Jones, P. 1978. *Community Education in Practice: A Review.* Oxford, UK: Social Evaluation Unit.

Krafick, John F. 1990. *Training and the Automobile: Informational Comparisons.* Washington, DC: Office of Technology Assessment.

Schaaf, Michael. 1977. *Cooperatives at the Crossroads: The Potential for a Major New Economic and Social Role.* Washington, DC: Exploratory Project for Economic Alternatives.

Sher, J. 1979. *Rural Education in Urbanized Nations: Issues and Innovation.* Boulder, CO: Westview.

Swack, Michael, and Donald Mason. 1987. Community Economic Development as a Strategy for Social Intervention. In *Social Intervention Strategies,* edited by E. Bennett. Lewiston, NY: Mellen.

Willis, R., and S. Rosen. 1970. Education and Self-Selection. *Journal of Political Economy* 87 (2): 350-366.

Wolman, Harold, Cary Lichtman, and Suzie Barnes. 1991. The Impact of Credentials, Skill Levels, Worker Training and the Motivation on Employment Outcomes: Sorting Out the Implications for Economic Development Policy. *Economic Development Quarterly* 5 (2): 140-151.

10 | Community-Based Economic and Employment Development

Community-based development initiatives are activities inspired by or aimed at serving particular social groups in a locality. In addition, community-based development initiatives also refer to those efforts organized by people who share a common urban geography. These efforts have proliferated in the last several years because of the failure of the general economy to serve the needs of particularly disadvantaged populations. These development initiatives aim to generate *socially useful, labor-intensive* projects that meet their expenses or make a profit while improving the employability of the participants. As Parzen and Kieschnick (1992, 5) put it so well:

> Ownership is integral to the notion of economic development. It has become commonplace to describe economic development by citing the old parable that you can feed someone for a day by giving them fish, or you can feed someone for life by teaching them to fish. This may be true. But the twentieth century postscript to the story is that what really matters is who owns the pond with the fish. There is surely a difference between two communities, one with all of its tangible assets owned by distant investors and one with a significant degree of local ownership. The absence of local ownership implies either a lack of local capital or a lack of confidence in local investing on the part of local owners of capital—neither is consistent with a policy of economic development.

The basic objectives of these initiatives, as identified in the relevant literature, are to teach people at the neighborhood level to own their own pond and to fish it. This is accomplished by doing the following:

- Generating employment for particular groups
- Gaining control over the local/neighborhood economy
- Inspiring self-help and cooperative group-oriented assistance
- Operating for the public benefit
- Providing an alternative or intermediate sector for economic activity
- Promoting democratic management and control of enterprises

Nearly all of the activities discussed elsewhere in this book can be undertaken as community-based initiatives. My experience, however, is that community initiatives are generally small and oriented toward the provision of household services, community services and facilities, and, occasionally, adjuncts to tourism or environmental programs. However, Teitz correctly warns all of us who espouse neighborhood or community economic development as follows:

> I argue that neighborhoods, by their very nature are problematic as a target of economic development strategy, insofar as a strategy is intended to generate economic activity and employment directly within their boundaries. As the places where people live and work, neighborhoods reflect most clearly the conditions of life of their residents. Thus, they are a logical focus of advocacy and political mobilization. But the historic divorce of workplace from residents is now so far advanced in American cities that localized economic development efforts face great difficulties. (Teitz 1989, 112)

In the face of such a caveat, communities must continue to press for some measure of control over the job-formation process and local economic activity. To accomplish this, suitable community-based organizations need to be established. The Clinton administration has pledged to give community-based initiatives new meaning as a component of national domestic economic policy. This chapter focuses on organizational form and objectives rather than specific activities.

Altogether, five organizational forms are discussed: community development corporations, community cooperatives, local enterprise agencies, employee/worker ownerships, and community employment and training boards. They are compared in Table 10.1.

Community Development Corporations

The community development corporation (CDC), an institutional form pioneered in the United States, is beginning to attract the interest of European governments.

CDCs were conceived as part of the War on Poverty in 1966, when Congress authorized a demonstration program as an amendment to the Economic Opportunity Act. The legislation, however, was vague, and the concept was still just beginning to take shape when the first CDCs started to receive federal funding in 1968. The early years of the CDC movement included a number of forays into ghetto development. Some basic principles that still guide community-based economic development were formulated then—community control, a comprehensive approach, a focus on business and economic development—but at that time no one knew how to apply them. Those early years also saw a number of failures, but even so, CDCs began to take hold.

They have gradually made the sometimes subtle transition from antipoverty agencies to economic development institutions. Their management and investment strategies have become more sophisticated. CDCs have expanded their resources, not so much through the original War on Poverty program or the later, now defunct, Community Services Administration as through other development programs, primarily in the federal government. To illustrate these changes, early CDCs concerned with community-control issues usually started and managed day-care centers, employment and training, and housing assistance, as in-house enterprises.

Few early CDCs considered investing in start-up businesses and few CDCs now manage their own ventures. (Examples are real estate development and technical assistance efforts.) Instead, CDCs tended to take minority equity and debt positions in proven (although often young) firms with experienced management. And although CDCs still select projects in response to community priorities, they do so within the constraints of market feasibility. Even so, CDC investments are risky, but in general they have become more successful with smaller investments. As a corollary to this trend, CDCs now work much more closely with the private sector, particularly small businesses and the financial community. Joint ventures with entrepreneurs, housing development syndications, and bank credit lines are now common CDC features.

The Senate Labor and Human Resources Committee, in assessing the effectiveness of various ways of delivering federal job-training programs, concluded that community-based agencies performed ex-

Table 10.1 Comparison of Organizational Forms for Community-Based Economic and Employment Development

Organizational	Objectives	Methods
Community development corporation	Build community-level institutions	Community organization and business formation
Community cooperative	Community/worker producer control	Collective business
Local enterprise agency	Unemployed/community business formation	Local resource mobilization
Employee/worker ownership	Worker control	Worker finance
Community employment and training board	Human resource development	Training

ceptionally well as economic institutions. Moreover, CDCs were considered cost-effective, which has encouraged the Ford Foundation to expend $9.5 million on a new Local Initiatives Support Corporation.

A long-term and comprehensive approach to development requires an institution that can develop and sustain a consistent but flexible strategy. CDCs provide full-time professional staffs and at least some planning capacity. Although some CDCs may rely heavily on community volunteers for help, most of their projects require technical skills applied intensively. They must be competent enough to develop communities avoided by most of the private sector. This formidable task requires the full energies of professionals. CDCs should, if possible, have at least some in-house planning capacity. This permits them to develop strategies that are coordinated, comprehensive, feasible, and responsive to community needs. Such sophisticated planning provides a considerable advantage over organizations that undertake only one kind of activity or cannot link overall strategies to specific projects.

Institutional Advantages of CDCs

Given the difficulty of developing low-income areas, the notion that community-based organizations have been able to succeed where others have failed or simply stayed away may seem difficult to accept. After all, where there is a market it should be adequately filled.

CDCs succeed because they consolidate issues and organize capital to deal with market failures in low-income areas. That is, CDCs reestablish the economic and social structure of low-income areas. Banks and other institutions only try to work on the economic issues without taking social barriers into consideration. Avis Vidal, who has completed the most comprehensive study of CDCs, concludes that they "make a major impact on the problems of poor neighborhoods" (Vidal 1992, 19). The financial success of a CDC project is due to sensitivity to both political and business practices. This necessarily demands a comprehensive and coordinated strategy. Because of their purpose, structure, and approach to economic development, CDCs combine characteristics that makes them unique institutions. They:

- use private development techniques for public purposes
- target benefits to communities and individuals in need
- mobilize local initiative to address local priorities
- take a long-term approach to development
- link planning to implementation
- link complementary projects within a comprehensive strategy
- understand and work with the processes of both the public and the private sectors
- legally can and in practice do attract both public and private resources in a variety of roles
- work directly with small businesses
- reinvest resources in the community
- have incentives to operate programs efficiently
- can transfer capacity among program activities

These characteristics are not, of course, limited to CDCs. Other institutions, including local governments, often perform some of these functions. Moreover, not all CDCs have the capacity to do all these things well. But CDCs are extremely flexible institutions. They can complement the activities of other local participants, develop their capacity through the assistance of these other institutions, and take increasing responsibility for development. In most communities, this process can be cooperative rather than competitive.

A city government, for example, may be able to use its powers of eminent domain to assemble land and its bonding authority to attract private capital for development, and then let a CDC develop and manage the project. In a rural employment training program, a local community college or the state's extension service might design a curriculum, while a CDC identifies and screens possible trainees and then places them with employers after the training is finished. States can contribute to these local partnerships by identifying roles for different organizations and by supporting innovative arrangements.

Community Reinvestment Programs

Community reinvestment is a social philosophy and a movement. Its aim is to replace capital that flows out of minority and disadvantaged communities by pressuring banks and other lending institutions to develop new lending practices for housing, businesses, and social institutions in low-income areas. The key to such reinvestment is the use of the 1977 Community Reinvestment Act, which states in part:

> that regulated financial institutions "have a continuing and affirmative obligation to help meet the credit needs of local communities in which they are chartered . . . consistent with safe and sound operation of such institutions." (Squires 1992, 11)

As a result, community reinvestment straddles an ideological dilemma of finding good loans in impoverished neighborhoods. Proponents of this movement have helped to devise a number of coalitions across the nation designed to identify neighborhood goods, such as housing, and match them with qualified minority or community-based institutional borrowers.

The community reinvestment movement has had some impact on reducing redlining and other adverse lending practices by major financial institutions. However, even its strongest proponents acknowledge that good investment opportunities that meet the test of the law are difficult to identify. Moreover, there remains enormous racism and selectivity in lending practices in the financial community. One response to the lack of community-based financing in minority communities has been the emphasis on the creation of nontraditional lending institutions such as community credit unions.

Community Cooperatives

Until recently, cooperative ownership was largely restricted to primary industry. But there is now growing interest in this form because of its labor-intensive character and its democratic management potential. Community cooperatives have been established in a number of states with some government assistance.

Several general situations give rise to the formation of community cooperatives, such as these:

- When a business owner wishes to sell to a community group because no other businessperson is available or interested (local newspapers provide exceptional opportunities for community cooperatives)
- When a community service such as child care, elder care, or other care requires delivery
- When a group of skilled people who are underemployed or unemployed form an organization to sell their services collectively for community benefit (for example, when a group of underemployed women open a regional clerical service in an area where no such services are available)

In each case, the steps for forming a cooperative are the same. First, a feasibility study needs to be conducted to ensure that the activity has real potential for success within a cooperative framework. It is important to understand that, in fact, very few businesses fit the this framework. Second, the organizational structure should be well thought out. Ownership shares, leadership, and similar matters need careful early discussion. Although there may be early enthusiasm for cooperative forms, this may quickly subside when people understand the system's rules. Finally, intensive training needs to be given to help the participants adapt to the new system.

Cooperatives require business plans as well as feasibility studies. Such work should at least provide answers to questions on the following topics:

Product Specification

- What evidence is there that a market exists for the product or service to be provided by the business?
- Are there similar products on the market or services available? If so, is there another local business offering the same product or service with whom you will be competing?

Workers

- Are the workers sufficiently trained in cooperative management?
- Is there real balance of skills among the workforce? Cooperatives will not work if there are differential skill levels.

Premises, Equipment, and Materials

- Are there low-cost premises available for the initial start-up?
- What transportation cost will be incurred in operating the venture?
- What are the sources of supply? Are they all under the cooperative's control? For example, is the cooperative selling products made by members or community groups?
- Can the cooperative obtain the necessary credit and access to capital to operate the facility?

Management Skills

Are there sufficient management skills in the group to provide for:

- designing the product or service?
- organizing the production or distribution?
- handling the banking and finance?
- marketing?

Community cooperatives are not easy to operate. They require an enormous amount of patience but can be a rewarding experience.

Local Enterprise Agencies

Local enterprise agencies (LEAs) are organizations dedicated to the creation of employment through the support and development of indigenous local enterprises, providing an intermediary link between public and private institutions and the community. They function primarily by facilitating access to capital and technical or professional assistance. LEAs are usually formed by a coalition of local business unions and government, along with community-based organizations. They facilitate nonprofit distribution that may or may not set up profit-making subsidiaries.

The primary functions of LEAs are to:

- Provide extensive advice and services for small business entrepreneurs through a permanent organization
- Provide complementary services that assist new businesses to get started by securing finance and technical assistance
- Build confidence in the total community by forming new networks of professionals and grassroots groups
- Improve local marketing capacity of new firms or existing firms

Employee/Worker Ownerships

Employee/worker ownerships are a recent phenomenon. They are a potential response to pending firm shutdowns. As companies decline, workers and community groups become interested in assessing whether or not they can be resuscitated by owning and operating them themselves. There are several legal structures that may be explored in this type of project. The two better-known examples are employee stock ownership and worker cooperatives.

Employee Stock Ownership Plans (ESOPs)

These plans allow a company to establish a trust in order for it to be purchased. The trust acts as fiduciary agent for the firm's purchase using stock issued to the employees. Employers purchase shares through a combination of wage reductions, actual cash, and borrowing. Over the years of the plan, employees acquire vesting rights to their stock. All stock owned by an employee is resold to the company in the event the employee leaves the firm or retires. The steps for forming an ESOP are as follows:

1. An employee corporation is formed.
2. The trust is formed and an ownership plan drafted.
3. ESOP borrows capital for equity contributions and operating capital.
4. Financial requirements are determined for the purchase of shares and to build equity accounts.
5. A debt service plan is organized.
6. The employee management structure is established.

State and federal laws govern the technical aspects of stock issues and trusts. These technicalities require close study by attorneys and other professionals because there are many potential pitfalls. Most ESOPs are likely to be in closely held rather than publicly held companies and provide an average of 10% to 40% of total company stock and limited voting rights, but the plans can be used to create majority employee-owned, democratically structured companies (Rosen 1989, 258).

Worker Cooperatives

Unlike the community cooperative, a worker cooperative is merely a business form. In worker cooperatives, the employees remain employees as such but are joint owners of the enterprise. Shares are issued to all of the employees and are reinvested for the firm's growth and development. Allocation of income at the end of the year is determined by weighted votes of employees, usually by years of service. That is, an employee with 3 years of service has more votes than a new employee.

A worker cooperative differs from a conventional business, however, in that membership rights are personal rights, assigned to the people working in the company because they work there. In other words, the workers have voting and profit rights in a cooperative because they work there, not because they have bought the rights. In a conventional business, by contrast, membership rights are transferable property rights attached to the shares of stock. The shares may be sold to any person who has the cash to buy them, and the person need not be connected with the business.

In worker cooperatives, the workers are the owners. The division of duties and responsibilities in such a system must be carefully considered. Not all employees may bring equal skills or talents to the firm. As a result, serious ruptures of the democratic management process may occur. Some members of the worker cooperative may not end up with a meaningful say in how firm development is carried out in their name. For this reason, staff development requires continual consideration.

Targeting and Marketing
Neighborhood/Community Assets

Marketing a community is very much like marketing a product. Product and market research are employed to determine what type of assets a community has to offer, in what markets, and to what type of clients. For maximum effectiveness, the community should be marketed only after determining the likely buyers of its resources, for example, tourism, where purchasers may be domestic or foreign. A community should develop a knowledge of its marketing audiences: What do they read? What appeals to them? How can they be reached? These questions are even more appropriate in industrial or commercial economic development. The market depends on whom the community is trying to reach. This is true for individual projects, attraction efforts, and financial assistance for local projects.

The fundamentals of targeted marketing are relatively straightforward. These are as follows:

Identification of the Community's Long-Term Aims

The level of long-term growth must be gauged with respect to overall population and/or jobs that a community is targeting. For example, in order to produce a job growth of 3% per year, an international audience may be the most likely target audience. Smaller, growing firms, however, may be more appropriate for a community that prefers more limited or controlled development.

A balance among different segments of industrial, commercial, and community-based activities may be desired as well. This balance determines what type of marketing efforts are undertaken and how vigorously they are pursued. In addition, a knowledge of the existing infrastructure will influence the development alternatives examined. In fact, it may be impossible to select certain targets without a long-range infrastructure plan.

One of the best examples of how a community has organized and projected its resource base is the emergence of Emeryville, California, as the nation's premier biotechnology center. Emeryville, with a population of less than 10,000, accomplished this by concentrating its infrastructure development on the requirements of the new biotech industries.

Presentation of the Community

Any presentation must depict the community as it is and the opportunities it presents. Glossy brochures and rhetoric will not help market a community. Major manufacturing firms have their own in-house research staff so they are not deceived by false advertising. Dissemination of misleading information will seriously damage an area's reputation.

The type of communication selected must appeal to the community's potential clients. If the prospective clients are Japanese businesses, information in one of the Japanese newspapers or magazines will reach this client base better than brochures written in English and hand-carried by the local mayor or community leader to the Japanese consulate.

Scientific or trade journals may be an even more important source for businesses than brochures or similar public relations efforts. If you want to get the attention of firms' managers, do it through the reading material of the senior executives. Whereas CEOs may read the *Wall Street Journal, The Economist,* or *The New York Times,* they seldom read brochures.

It is important to know the industry being recruited. Neighborhood or community leaders need to know some general details about the type of firm likely to visit the community. Many industrialists or would-be entrepreneurs have been discouraged by the seeming lack of interest or shallow responses of community leaders to questions regarding the industry and its requirements.

Incentives

Incentives should be packaged to meet the needs of the firm or individual entrepreneur a community is trying to attract or encourage in the area. Financial incentives are seldom as important as a good business climate, available research facilities, appropriate space at reasonable rates, and an aggressive local government with a flexible approach to incentives.

A community should list available assistance and resources as well as location information. Communities often publish promotional material with no information on whom to contact for further details. As a result, they have missed sales simply because of poor communication.

The entire community should be involved in promotion. (There are many stories about uncooperative staff giving a negative impression to visitors in casual conversations.)

The Community Prospectus:
A Means of Communicating the
Qualities of Your Community

Many communities invest in general promotional brochures. Although these may be helpful, they seldom attract a firm. Because targeted marketing requires establishing the specific feature identification required by an industry or certain type of firm, the best means of communicating to firms is through the community prospectus. This material acts much like a firm prospectus, and it allows the buyer to know what the community is offering and what opportunities are available.

The prospectus should describe the community assets and business opportunities in terms of industrial and market potentials, as well as community trends over the last few years and expected future trends in commercial and industrial opportunities. It should also relate any national economic or social trends that may affect a venture's potential in the community. Moreover, it should include a section on the community's research and development capacity, infrastructure plan, and its financial assistance program.

The *research and development capacity* section is one of the most important components of any community prospectus. Access to higher education institutions, research laboratories, or facilities offering scientific knowledge and/or trained personnel is vital to almost all industries. It is important to describe the nature and extent of the scientific efforts in the area that relate to business, industrial, or agricultural products and services.

In addition, other private sector research or advanced facilities or equipment should be described as well as any major products, scientific breakthroughs, or well-known researchers resident in the area. Of course, very few communities have such resources within their boundaries, and most communities would benefit from projects that improve their research and development capacity in defined areas that complement their development strategies.

The *community infrastructure plan* section of the community prospectus should discuss the community physical infrastructure—location and buildings—specialized equipment, and personnel (other than management) available to prospective businesses. It should describe what the community's industrial or retail/commercial space availability is in terms of specific venture requirements. In order for this to be effective, suitable building sites should already have been located and fully developed. If a site has been selected, describe it and how it meets the venture's needs, the businesses surrounding the site, whether the site and any buildings on it will be bought or leased, and the reasons behind these decisions.

The infrastructure discussion will cover slightly different topics depending on whether the venture is manufacturing, retail, or service. Existing plans should be examined for the targeted businesses. Infrastructure descriptions for both retail and manufacturing should list the availability of water, power, and waste removal, as well as any environmental constraints and existing transportation systems. Telecommunications is an especially important infrastructure component and it needs to be carefully assessed as a significant infrastructure requirement. Today optic cable, international switching, and data transmission facilities such as teleports are as important as water and sewer capacity. Manufacturing discussions should also address the availability of industrial space at reasonable rates and whether facilities are available on a short- or long-term basis.

The description of human resource availability should include what kind of labor can be obtained locally for the type of venture(s) sought. This section should also discuss the executive and managerial skills available in the area, as well as the general labor supply. If training is required, there should be a discussion on how it will occur and the associated costs.

The *financial program* section of the community prospectus is the last section because it is based on the information in the earlier sections. It should offer plans to improve the critical gaps in the community infrastructure as well as offer specific financial assistance for the desired type of firm. This section should, of course, make note of the financial programs available; these include initial financial assistance, venture capital, debt financing, and other specialized funding. Additional community funding or special arrangements for access to local capital should also be described.

Local Government's Role in Community-Based Initiatives

Local governments can take active roles in stimulating and supporting community-based initiatives by doing the following:

- Developing revolving loan funds and similar financing as seed capital for community projects
- Examining means to subcontract to community groups the delivery of community services or the operation of community facilities
- Developing work spaces and facilities for community groups to commence enterprise activities
- Identifying surplus or underutilized local government equipment that might be used by community groups
- Establishing a network of people or service clubs that can provide technical assistance to community groups
- Making community-based initiatives a component of the overall local economic development strategy

---

Illustration 10.1. The Greenhouse Venture

An organization may take an equity approach to CED but choose to limit its initial activities to small ventures that are closely related to existing activities that utilize existing staff and fiscal systems, that require little capital investment, and that pose only limited risk. These ventures are called "greenhouse" ventures. Pursuing small-scale greenhouse ventures can provide an organization with valuable developmental experience, while posing little risk of failure. If the greenhouse venture is successful, it can be expanded and eventually spun off as a free-standing venture. Some successful ventures in the Law Center's technical assistance program began as green-house ventures. Although small in size, these ventures generated significant benefits for the organizations that pursued them.

For example, by 1986, when it joined the Law Center's technical assistance program in Seattle, Aradia Women's Health Clinic was suffering from the fiscal woes of many community health clinics that serve low- and moderate-income women on a sliding fee basis. Already operating in the red, the Clinic felt that it could not commit a significant amount of staff time to CED or risk any development capital. It searched for a greenhouse venture closely related to its existing mission, one that would not require significant staff time or development capital, that would not require major changes in management and fiscal systems, and that would quickly generate a profit.

As a women's health care provider, Aradia recognized that women constitute a growing percentage of the potential purchasers of condoms. Every day, dozens of women passed through the Clinic's reception area, women who were potential purchasers of condoms and other contraceptive supplies. The Clinic could appeal to this market by offering women a diversity of product lines and a supportive environment in which to consider a purchase. Aradia established a display case for condoms in its reception area and began selling them to clients.

Aradia's greenhouse venture required no new staff, almost no initial capital, and only modest management and fiscal adjustments. Moreover, it was entirely consistent with the organization's social goals. Within only a few months, the venture succeeded in generating net income of more than $500 per month, cutting the organization's operating deficit almost in half.

More important than the profits that the greenhouse venture generated, however, was the organizational change it promoted. The members of the Aradia collective began to look at the clinic as a business, rather than a social service program. Aradia developed a successful advertising and marketing campaign for clinic services. It investigated the cost and methods of financing special equipment that would enable the Clinic to significantly increase its billings to third-party payers. And it examined the feasibility of controlling its space costs by acquiring its own building.

With the understanding that CED encompasses several different kinds of organizational commitment, an organization's decision to say "Yes" or "No" to CED is more informed. Whether pursuing a facilitative or an equity approach to community economic development, a CBO [community-based organization] must begin with a careful assessment of its organizational strengths and weaknesses, assets and liabilities.

SOURCE: Wiley 1989. Reprinted with permission.

Illustration 10.2. Colorful Patchwork

One of the Tutwiler clinic's most successful efforts is the sale of brightly colored African-American quilts and quilted bags made by local women.

The project began in 1988—the brainchild of Sister Maureen Delaney, outreach director at the clinic. Sister Maureen is a community organizer from Oakland, California, who has been living in Tutwiler and working on outreach projects for five years.

Fifty local women are involved in the quilting effort, which allows them to earn money while contributing to the clinic. Prices range from $170 for a twin-sized quilt to $250 for a queen-sized model. Quilted bags—which have been sold at the Smithsonian Institute in Washington, D.C.—go for $20 each.

"The women get 70 percent of the price of the quilts," said Sister Maureen. "Nineteen percent goes toward buying material and eleven percent to supporting the outreach van program."

"It lets me make a little money and I'm proud of that," said Lady B. Lloyd, who lost a leg to diabetes and pieces together quilt covers in her apartment across the street from the clinic. "I just love to do it. Most of the time I sit here all day by myself and I piece them together."

Williebelle Schegog, another Tutwiler resident, said she works on quilts from 9 a.m. to 3 p.m. every day except Saturday and Sunday. "Those are the Lord's days," she said.

"Quilting gives us something to do," added Alberta Mitchell, a member of the Tutwiler Improvement Association who also serves on the Community Education Center board.

Today, hundreds of mail orders are waiting to be filled, thanks to exposure the quilting program received on a CBS "60 Minutes" broadcast. There's a backlog of about nine months, according to Lucinda Berryhill, chairwoman of the Tallahatchie County School Board, and coordinator of literacy and quilting at the clinic.

The national television exposure also unleashed a flood of donations to the clinic's two Bargain Barn stores, which offer low-cost clothing and other useful items to area residents.

SOURCE: Fuster 1993. Reprinted with permission.

Illustration 10.3. Neighborhoods and the Urban Crisis

The scope of inner-city troubles—the joblessness, destabilized families and decaying institutions—is so enormous that small-scale rehabilitation projects seem dwarfed by them. In the absence of federal economic policies that either create or stimulate jobs for inner-city residents, it may well be that cities are doomed to continue to suffer. And yet even the most skeptical observer has to notice that the neighborhood approach is quietly piling up evidence of change for the better.

One of the best pieces of testimony comes from Miami, which has suffered through several riots, most notably the 1980 Liberty City conflagration that

left 18 people dead and parts of the neighborhood in ruins. In its aftermath, the city got off to a false start. Its business leadership decided that the answer to the city's problems lay in unleashing the private sector, and it set out to stoke black entrepreneurship. It wasn't very long, though, before the business community retreated, frustrated by the high failure rate among the small enterprises it tried to help and community resistance to the outsiders it brought in as advisors.

Meanwhile, with little fanfare, a handful of local officials and others put together a package to help a neighborhood group, the Tacolcy Economic

Development Corporation, buy a burnt-out shopping center. It took five years of patching together federal, state, local, and private money for Tacolcy to redevelop the property and lure a tenant, but the Winn-Dixie grocery chain finally opened a store there.

In the scale of things, it was a small victory. But it also proved to be a catalyst. The supermarket sparked renewed interest in the area, and a drug store and a photo processing shop opened up next door, a McDonald's went in nearby, and storefronts all along the street were spruced up. The city helped out by putting in brick sidewalks, new street lamps and, perhaps most importantly, a police substation. The community college has opened a branch nearby, with an entrepreneurship center to help black-owned businesses.

"It is looked on as one of the few urban success stories," says Ernie Martin, who recently retired after 14 years as community development director for Dade County. "It produced an optimism that brought in franchises and chains, it produced an environment in which banks are willing to lend to black-owned businesses, and it produced housing that's attractive to young people who would otherwise have moved out. It has created an environment that the business community tried to create, but couldn't do from above."

Something similar has happened in Pittsburgh's beat-up Homewood section, a neighborhood to the northeast of downtown, where the Homewood-Brushton Revitalization and Development Corporation has turned an empty boarded-up central street into a revitalized commercial corridor. Starting in 1984 with a corner that it bought for $300,000—pieced together with money from the city, a foundation and a couple of banks—the group wooed a Dairy Queen franchise and then an Athlete's Foot and other shops; now it is working on a $4.5 million mini-mall. Homewood-Brushton publishes community newspapers, runs a 1,000 watt radio station, and has been developing low-income housing units. In all, it controls a hefty chunk of the commercial space along the six-block strip off Homewood Avenue, giving it the ability to lure minority businesses with low rents and the promise of an exclusive franchise on the blocks it owns.

What the group is doing, says the man primarily responsible for the turnaround, is more than simple economic development: it is the re-creation of a community. Mulugetta Birru, once an official in the Ethiopian government's Agricultural and Industrial Bank, ran Homewood-Brushton from its founding until recently, when he was made director of the Pittsburgh Urban Redevelopment Agency. "You don't want your community to be just a place people sleep," says Birru. "You want them to be involved in it." Commercial development, he says, "creates a sense of a village environment, where goods and services are available and people meet and talk while shopping." By the same token, the radio station has become a kind of community bulletin board, over which listeners share information—including, for instance, sites where drug dealers have been seen trying to recruit help.

SOURCE: Gurwitt 1992. Reprinted with permission, *Governing Magazine,* copyright © 1992.

Case Study 10.1. Workers as Owners

On a late autumn evening in northeast Franklin, a bookkeeper and a couple of stock clerks could be seen through the plate-glass windows of the ComUNITY Sporting Goods Store. For everyone else, work had ended. But for ComUNITY Sports' directors and stockholders, sitting on empty running shoe boxes and sports equipment crates, the day had only begun. The president, Joe Offer, called the meeting to order, and soon the group was engrossed in a spirited debate on product mix. Some people advocated cutting the price of ski equipment early in order to stimulate sales in all of the city's ComUNITY Sports outlets. This had been a very successful tactic of the old management. Several people objected to this proposal simply because it was the old management's idea. "Who wants to be associated with a management that was forced into near bankruptcy on a nationwide sporting goods chain?" Another hot topic was corporate directors' liability. Both topics were being pursued in heated debate simultaneously as Joe Offer, the recently elected CEO and stores manager, called the group back to order.

At this point, Jim Schader, then board secretary, spoke up. "Dunn & Bradstreet's asking for details on our financial status," he said. "What's our policy on giving out this type of data to investment services?"

No policy existed, but there were plenty of opinions both on the point and at some distance from the actual task at hand. When no consensus was forthcoming, Offer finally expressed what turned out to be the majority view. "Our credit's good," he said, "and we will provide it because it will be good for us to cooperate early in order to get the best credit rating possible when we need it. We are directors and the only stockholders of the company."

And an unusual company it is. Until last year, the store in which this meeting was held and 10 others in the city of Franklin were combined with a chain of over 200 discount sporting goods stores. The company had prospered in the 1980s when discount consumerism was rampant, but the whole chain was in danger of collapsing by the time the employees made an offer to take over the entire West Coast division of the company and break it up into smaller city-based retail outlets separately managed by the workers at each store but under an umbrella corporate structure. Today, the ComUNITY Sports stores claim that weekly sales are 20% above those of other retail clothing and equipment chains: The Roslyn store sales are up 40%.

"It's *our* store," says Faith Mason, ski sales manager for all of the ComUNITY Sports stores. "We all feel that way about it. Owning the company gives us more incentive to make it a success. We're putting in about 80 hours a week—40 as workers, 40 as owners. And guess what? I actually *enjoy* coming to work. There will be more store systems like this one. It's a great program and too good an idea not to spread."

Yet, support for employee ownership was not universal at the time it was proposed. In Franklin, for example, Cornell Young, a labor leader, encountered criticism when he first explored worker ownership a few years ago. "An executive at ComUNITY once told me it sounded like socialism to him," Young says, "but I said we just wanted to organize stores the way he and his friends had originally started back in the late '60s when they were radicals at Berkeley—everybody pays into a kitty, you elect some directors, you hire a manager. What's socialistic about that? That's exactly what they did before they joined the country club set."

Young has served for years as president of Local 357 of the United Retail Sales Union (URSU). The local's membership peaked at about 9,000 in the mid-1970s, after which the depressed economy forced a string of sporting goods retail stores to close, which cost the union more than a quarter of its members. The crowning blow was when Pacific Tex, the conglomerate that spun off ComUNITY as part of its profit center in 1977, decided in 1981 and 1982 to shut down 14 of its West Coast-area stores, idling nearly 400 URSU workers as well as members of a smaller textile union. Pacific Tex had closed more than 150 stores in eight years—65% of its nationwide chain—and its losses in 1981 exceeded $30 million.

In many respects, it is a unique and unprecedented agreement. Workers obtain ownership of a segment of the company and the right to buy in the future as well as more authority and responsibility. It provides an avenue to learn management skills and provides a legal property claim; a right of first refusal to purchase stores from the company, so it cannot be bid up competitively and sold off to someone else. This arrangement creates an important source of worker-controlled capital—perhaps as much as $1.5 million a year at first, with more as time goes on—to finance incentives and investments in employee-owned enterprises.

Pacific Tex officials acknowledge that their main interest was and is in the stores they kept, not the ones they gave up. Without making judgments on the sales potential of the stores, they do characterize them as "experimental." According to one company official, the agreement reflects "a sharing of objectives with the union. U.S. business and communities have to find ways of working together and stop fighting with one another; that's the only way we can be internationally competitive."

Even while this experiment is under way, the directors of ComUNITY need to consider diversifying beyond sporting goods into men's and women's clothing lines imported from China. Since the union has gone on record as "Buy America" this is not an easy decision. But if a move is not made soon, the managers fear the ComUNITY experiment will disappear forever. This is a tough issue. The employer/owners feel that selling foreign goods will hurt local workers.

The board doesn't know how to raise this issue with the membership or the impact of failing to compete in a global marketing system.

The board has called you and your team as consultants and they want your advice on how to keep their experiment going.

Case Study 10.2. A Community Finds a Business

Lakes Alternative Energy Board (LAEB), a community development corporation, is composed of four community action agencies involved in alternative sources of energy. This partnership will simultaneously bring in revenue to the community, create jobs, and provide low-cost fuel to area residents.

LAEB serves the Lakes area, which is known for its high unemployment rate and high percentages of elderly residents and families with incomes below the poverty line. The area has been severely hit by the collapse of the two traditional sources of employment and revenue: the iron mines and the lumber industry. Fuel is also a primary concern in the region, with its 120-inch annual snowfall and temperatures below freezing for 2 months of the year. Fuel costs are high; many of the Lakes's elderly clients spend about half their income on fuel.

The first venture is the use of wood pellets as a fuel source, which will eventually lead to the building of a wood-pellet processing plant in one of the nine counties served by the four agencies. This mutually beneficial venture illustrates how community action agencies can turn their nonmonetary assets into an effective base for the creation of community economic development projects.

Another project was to have been the construction of live-in demonstration models of energy-efficient houses: retrofitting existing housing, developing wood-based insulation materials, and producing low-cost stoves. At the time, though, the energy crunch was not a hot topic and the proposal was turned down.

LAEB has been incorporated by the three CAAs (community action agencies). It is investigating new, alternative energy sources with economic development potential.

The goals of LAEB included analyzing the use of wood refuse and small streams on farm property as an alternative fuel source, especially for low-income people; promoting the use of alternative fuel sources; and creating—and keeping—jobs in the wood industry through the private sector. The intent is to develop a low-cost fuel that can be used by both commercial and residential energy systems.

LAEB has stressed to Lakes residents that it is not in competition with the private sector; rather, it sees its function as a catalyst for local economic development. LAEB staff members define their role as helping private entrepreneurs get established in the alternative energy market. Possible spinoff ventures include maintaining storage facilities for bulk quantities of pellets, manufacturing the special stoves needed to burn the pellets, and running local dealerships to sell and service the stoves. LAEB also wants to establish a market for wood refuse that would generate income for loggers, truckers, and the struggling wood-processing plants in the area.

LAEB is attempting to involve all sectors of the community in its efforts. Representatives from the four community action agencies sit on the board, along with representatives from the lumber industry, local universities, the U.S. Forest Service, and cooperative extension services in each county, as well as low-income people from the area. There are also a number of technical advisory groups and committees set up to bring in authorities on various aspects of the lumber industry.

Once LAEB established its board and narrowed its focus to the development of wood pellets, it turned its attention to finding a private partner with the necessary experience to establish a pellet plant. Asperal, Inc., a wood-pellet manufacturer in Minnesota, approached LAEB when it heard that the board had capital to lend to a private partner in exchange for establishing a pellet plant.

LAEB evaluated the quality of Asperal's pellets and its collateral, and then negotiated a loan agreement with the company. In exchange for a low-interest loan, Asperal will work to establish a market for wood pellets in the Upper Peninsula. If successful, Asperal will begin building a pellet plant in one of the nine counties covered by LAEB.

Capitalizing on their status as nonprofit entities and their access to federal funds, the three community action agencies in LAEB used a government grant to cement a public/private partnership with Asperal. As the private partner, Asperal, Inc., benefits from the agencies' contacts in and knowledge of various sectors of their communities in gaining the support and cooperation of area residents. LAEB gains the business experience and technical knowledge of the private firm that will help ensure the venture's success.

The question before LAEB is what its next venture should be. The energy crisis is over and pellet sales will lag soon. How can all the energy and expertise be directed to creating jobs in the future?

Case Study 10.3. New Freedom House

New Freedom House was founded to set up a Native American jewelry distribution center on the Napahmoe Reservation in Taos, New Mexico. It markets products for four American Indian cooperatives (co-ops)— three that produce leathercraft and one that makes garments. The co-ops were the first established by the National Center for Native American Enterprise (NCNAE).

The initial distribution business via national sales catalog did well. People responded favorably to the products (design and quality) and the Christmas season brought substantial sales. After the holidays, however, sales slowed considerably. Freedom House responded by entering the walk-in retail business in Santa Fe, while continuing as a mail-order operation. The business also bought raw materials for all the worker co-ops, many of whom had been buying their supplies at excessive prices or were simply unable to purchase supplies in their local areas.

It was then, too, that NCNAE decided to turn over ownership of New Freedom House to the reservation tribes who supplied it with its merchandise and to those who worked in the various co-ops. The resulting cooperative would market goods for individual craft co-ops, buy their supplies, and have as members all those who worked in established co-ops.

New Freedom House Cooperative (NFHC) is a two-tiered organization. The first tier consists of the "unit members," or unincorporated associations of individuals organized on a cooperative basis who are engaged in the production and sale of handcrafted products. There are 13 unit members, all situated in Colorado and New Mexico, including, in addition to jewelry, several leather co-ops, a clothing co-op, and a candle co-op. Each unit member is entitled to elect one representative (usually its president) to NFHC's board of directors. The board elects officers who in turn employ professional managers and a staff to operate the main office and warehouse.

Each unit member (workshop) is a separate entity, but all are organized on identical lines and their individual members sign identical Articles of Association. Any person may become a member of an association at any time by subscribing to its articles and paying dues. The members elect officers who function as a management committee. The number of partners ranges from 6 to approximately 25.

The By-Laws and Marketing Agreement signed by each unit member are the two basic governing instruments describing the various responsibilities of the members and the directors and officers.

Each workers' co-op (unit member) is responsible for the production of particular items in the product line described in the Freedom House catalogue. These products include suede handbags, leather belts, cuff links, costume jewelry, dolls, stuffed animals and toys, children's clothing, canvas bags, candles, greeting cards, and numerous other items. Each workshop, moreover, is responsible for maintaining its own records of its total production and that of individual members. Every 2 weeks, an employee from the main office of New Freedom House collects the finished products and delivers raw materials and equipment that unit members have ordered.

Partial payment in the form of an advance is made for the products on the basis of a cost price that includes labor costs, costs of material, and overhead items for each unit member. The advance is distributed to workers on the basis of their weekly production.

After being collected from individual workshops, products are taken to the New Freedom House warehouse, where each item is tagged and inventoried by quality-control personnel. Any item not conforming to established standards is tagged for return to the particular workshop. After clearing quality control, all of the products in a particular category are pooled and placed in storage bins or on racks for distribution through retail outlets or mail-order sales on a national level. Proceeds from mail-order sales account for most Freedom House income. Mail-order catalogues are distributed to an extensive mailing list throughout the United States, Canada, and England. This list is constantly augmented by trading with other organizations and by leasing lists from direct-mail brokerage services.

NCNAE has organized several urban-based organizations, including six retail outlets plus Graphic Arts Southwest, Southwest Media, and Flute Publications. The latter are affiliated as craft workshops—that is, the publications and media workers are members of NCNAE. Although their activities are conducted in an urban setting, their activities are directly related to the needs of the rural members.

New Freedom House has faced the future squarely and the prospects are not good. The organization is large and too decentralized to compete with the better-capitalized mail-order firms tied in with the credit card companies. But New Freedom House has both a mission and a message. The board is now struggling with a proposal from a large New York mail-order firm to take over management and distribution of its products. Board President Chief Joseph reminds his fellow board members that neither the Anglos nor any other minorities were there when they needed them in the early days. Young Susan Windfall, another board member, says the Chief's ideas are tired and racist. She believes it is time for Native Americans to show the way—and make big profits. Who is right? How would you go about setting up an agreement with an external retailer to work with the Native American community? Or would you reject the idea altogether and redesign the firm? If the latter decision is made, what would you do?

Summary and Conclusion

Community-based economic development and employment initiatives are important elements in a wider local economic development strategy. In general, these initiatives have strong social objectives underpinning economic ones. They are as much rooted in and motivated by needs (of groups, individuals, and localities) as they are in finding and exploiting economic opportunities. They offer a structure and resources to encourage economic activity at the neighborhood level. They provide access to job opportunities for local people who find entering the economic system difficult. (See Table 10.2 for a sampling of community-based employment development projects.)

The track record of community-based economic and employment development organizations has not always been good. Some of the reasons for this have been an imbalance in the mix of social and economic objectives, inadequate financing, and lack of skills and planning. Another major problem relates to the fundamental issue of ownership and control. In theory, community-based orga-

Table 10.2 Community-Based Employment Development Project Idea
            Starters

Tool	Project Idea Starter
Employee/ownership cooperatives	*Skilled trades cooperative:* A cooperative is formed of skilled people of various trades who offer advice on technical matters to local firms and provide a sales and repair service for all kinds of specialized equipment and machinery. *Worker acquisition of existing firm:* Industry workers agree to put up money of their own—and accept a "readjustment in their pay"—in a bid to save their jobs and buy a sizable share in their former employer's business through the formation of a new company owned jointly by workers and perhaps a state equity bank (e.g., metal workers with the blessing of a union and the support of a new equity bank). *Cooperative government projects bidding service:* People with expertise in various aspects of fabrication come together as a cooperative to bid collectively for contracts (government and other) that they do not have the capacity to bid for as individuals.
Community-based employment	*A home-energy auditing and retrofitting scheme:* The aim is to provide an energy auditing and efficiency test service to home owners and prospective buyers to estimate energy costs and to provide quotes on retrofitting to reduce costs. The project also might supply a team of people capable of carrying out retrofitting activities.
Community-based service and employment development	*Recycling of waste as art and craft materials:* This project is to collect and redistribute industrial and other waste as art and craft materials from a shop/center that may be used by anyone in the community who pays a yearly subscription or a set fee per garbage-bagful. Also, lessons might be provided in making art and crafts with recycled materials.

nizations for economic development and employment generation are democratically owned and controlled. In practice, however, problems have occurred when these organizations commence projects on a business basis. Finally, as Michael Teitz (1989) warns, these problems are mentioned here not to undermine the concept but to state the limitations clearly.

References and Suggested Reading

Birch, D. 1979. *The Job Generation Process.* Cambridge: MIT Program on Neighborhood and Regional Change.

Bradford, Calvin et al. n.d. *Structural Disinvestment: A Problem in Search of a Policy* (mimeo). Evanston, IL: Northwestern University for Urban Affairs.

Carlson, David, and Arabella Martinez. 1988. *The Economics of Community Change.* Study funded by the Ford and Hewlett Foundations.

Cummings, C., and N. Glaser. 1983. An Examination of the Perceived Effectiveness of Community Development Corporations: A Pilot Study. *Journal of Urban Affairs* 5(4).

Daniels, B., N. Barbe, and B. Siegel. 1981. Experience and Potentials for Community-Based Development. In *Expanding the Opportunity to Produce,* edited by B. Daniels. Washington, DC: Corporation for Enterprise Development.

Daniels, B., and C. Tilly. 1981. Community Economic Development: Seven Guiding Principles. In *Resources,* edited by B. Daniels. Washington, DC: Congress for Community Economic Development.

ESOP Publications and Materials. The ESOP Association, 1100-17th St., Washington, DC 20036.

Fuster, Stephen Collins. 1993. *WKKF International Journal* 3(1). (Battle Creek, MI: The Kellogg Foundation)

Gurwitt, Rob. 1992. Neighborhoods and the Urban Crisis. *Governing Magazine* 5(13).

Haberfeld, Steven. 1981. Economic Planning in Economically Distressed Communities: The Need to Take a Partisan Perspective. *Economic Development and Law Center Report* (December).

Hein, B. 1987. *Strategic Planning for Community Economic Development.* Ames, IA: Iowa State University Extension.

Kotler, Milton. 1971. The Politics of Community Economic Development. *Law and Contemporary Society* 36(Winter).

Parzen, Julia, and Michael Kieschnick. 1992. *Credit Where It's Due.* Philadelphia, PA: Temple University Press.

Pierce, Neal, and Carol Steinbach. 1990. *Enterprising Communities.* Washington, DC: Council for Community-Based Development.

Rosen, Corey. 1989. Employee Ownership: Promises, Performance and Prospects. *Economic Development Quarterly* 3 (3): 258-265.

Snipp, Matthew C. 1988. *Public Policy Impacts on American Indian Economic Development.* Albuquerque, NM: Native American Studies Institute, University of New Mexico

Squires, Gregory, ed. 1992. *From Redlining to Reinvestment.* Philadelphia, PA: Temple University Press.

Teitz, Michael. 1989. Neighborhood Economics: Local Communities and Regional Markets. *Economic Development Quarterly* 3 (2): 111-122.

Vidal, Avis. 1992. *Rebuilding Communities: A National Study of Urban Community Development Corporations.* New York: The New School of Social Research, Graduate School of Management and Urban Policy.

Wiley, Jean. 1989. Greenhouse Ventures. *Economic Development & Law Report* 18(1).

Yin, Robert, and D. Yates. 1975. *Street Level Governments.* Lexington, MA: Lexington.

11 | Preparing a Detailed Project Plan

Why the public is investing in private sector activities is the central question researchers and practitioners should be asking (Hill and Shelley 1990). The answer lies in the complexities of revitalizing areas neglected by the private sector. In response to this neglect and/or market failures, the public sector needs to stimulate private lending. It also needs to provide the security and risk protection investors require to balance publicly desirable investments with those that are demonstrably profitable (Bingham, Hill, and White 1990).

Public-private partnerships are a long, difficult, and time-consuming process. First, the project idea must be determined to be viable within the environment it is designed to improve. Second, these projects must be shaped to pass the viability test by building action plans specifying in some detail the project, outputs, resource, and support system requirements. For smaller, simpler, shorter-term projects, an action plan may itself be adequate to gain formal commitment for the project idea and thus claim materials, finance, and personnel to implement it.

As projects become larger, more complex, and longer-term, however, action plans for the project design need to be fully detailed to claim resources. Detailed project plans are also required whenever a local government is considering acting on its own, or if projects are submitted that invite local government endorsement.

The preparation of a detailed project plan, which constitutes the fifth phase of the local economic development process, has four basic tasks associated with it. They are the following:

1. Viability assessment
2. Detailed feasibility studies

3. Final design and business plan preparation
4. Design of a monitoring and evaluation program

The main difference between the detailed project plan and the formerly described action plan is the degree of detail required. I concentrate in this chapter on general standards for detailed project plans that can be adapted to any requirement.

Assessing Project Viability

Project development is the redefinition and determination of the potential of specific project ideas. During this stage, the decision must be made whether to complete the planning phase and implement the project. In general, project development involves taking one or more attractive project ideas, shaping and specifying each precisely, and rejecting those that do not seem viable. To justify the project, the criteria of potential for economic development and employment have to be satisfied, at least preliminarily. In project development, viability is determined in relation to four interconnected bases: community, locational, commercial, and implementation (Malizia 1985).

Community Viability

Because economic development is a community issue, the formulation and selection of local economic development projects constitute a community-wide political act. First, the local economic development process describes the necessary support base required to make a project viable. Then it decides whether or not the project has sufficient community support.

Caution is suggested when deciding whether a project has community viability. Many individuals and organizations may initially give casual support to project ideas but then withdraw their support when the project is likely to become a reality. To overcome this potential problem, local economic development planners should seek explicit oral commitments from other key actors. These informal commitments are required because to proceed on any other basis is far too cumbersome. Formal commitments are sought only after the project has been found to be generally viable, an action plan assembled, and all of the important actors and their roles identified.

Locational Viability

For the locational/market viability analysis, the resources required by the project need to be identified and their availability determined. How the proposal harmonizes with other existing or proposed projects needs to be considered. Next, planners should identify potential customers for the product or service provided by the project and answer the following questions: Who are they? Where are they located? Why would they buy this product? How frequently? Planners then need to step back and consider market size and trends. They should determine if the product is in a growth industry, if the market area is growing, and what the expected size of the market is when operations begin. The next step is analyzing the competition. What are the strengths and weaknesses of the product relative to the competition? What is the market capture strategy, that is, what will be done to gain customers?

The answers to the above questions will provide two valuable pieces of information: (1) whether to proceed any further with the project, and (2) what should be included in a formal market study and marketing plan by consultants in a later stage of the economic development process. If the project appears to pass the test of locational/market viability, its economic/financial and operational viability then needs to be considered.

Commercial Viability

Based on the answers to the questions asked in the market viability analysis, the community or other sponsor must estimate the revenue-raising potential of the project and judge the likelihood of the project recovering its costs. Knowing the level of risk to capital is a precondition to determining whether investors will be readily available and whether the project will have reinvestment opportunities. The economic viability studies should consider each of these aspects. They constitute the initial pieces of a project cost plan to be made in the next step of the local economic development process.

Implementation Viability

Implementation viability is determined by considering who in the community has the skills and capacity to undertake the task.

There is no real sense in embarking on projects if the skills to organize or manage the project are not present. In addition, the community has to consider how the services and goods are produced and the means of production, as well as whether the Five M's can be satisfied in developing and operating the project. On completion of this task, the financial feasibility must be determined before the project is undertaken.

Detailed Feasibility Studies

The purpose of the detailed feasibility study is to identify the critical factors affecting each project's potential for success, to specify what conditions or requirements are necessary, and to evaluate the possibility of achieving these requirements. The information previously gathered for the project-viability analysis and the action plan forms the basis of the full-scale feasibility study. This study investigates in more depth any weaknesses identified in the earlier steps and does a more thorough analysis under each feasibility factor—especially the financial aspect. The research should result in an internal working document, usually 15 to 20 pages long, which in turn forms the basis of the business plan. The business plan is a more formal document, designed for distribution outside the project planning committee. It incorporates much of the information in the feasibility study but also discusses how the conditions for success will be achieved vis-à-vis the operation of the business.

There is no standard format for a feasibility study, but there are a minimum of three special studies that should be undertaken:

1. Market analysis
2. Financial analysis
3. Cost-benefit assessment

These studies should be undertaken for project start-ups, business expansions and acquisitions, and community development projects. Research and evaluation methods, however, may differ somewhat in each case. The scope of each study will depend on the size and complexity of the prospective venture. If any major problems or weaknesses are identified, they must be examined thoroughly, increasing the study's length and complexity. Although the precise nature of each of these analyses will differ according to the venture under investigation, certain factors should be addressed in each of them. These are discussed below, and sources for gathering information will be given throughout.

Market Analysis

At earlier stages in the planning process, I looked briefly at the overall market for a project's potential products or services. I also roughly defined the project's potential customers and estimated their numbers, based on general research. The market analysis component of the feasibility study is more specific and detailed. It defines the project's primary and secondary markets and determines the total size and share of the market that the venture can be expected to capture. To do this, the market analysis must closely examine three factors: the project's products, the total potential market, and competition.

Products

The first step in assessing the feasibility of all possible projects is to specifically define the product or service, or the mixture of goods and services, that the activity or business will produce and/or sell. As discussed earlier, all other factors in the feasibility analysis will relate to this definition. If the venture will manufacture a product, this part of the analysis may include a technical study, covering topics such as these:

- Crucial technical specifications, such as design, durability, and standardization
- Engineering requirements, such as necessary machines, tools, and work flow
- Product development, such as laboratory and field test results or plans

If the community does not have the capacity either to pay for or to carry out a technical product analysis, the venture is not a feasible one for that organization. If the venture will sell, but not manufacture, the product, as in retail or wholesale undertakings, the product analysis may consist of comparing products currently available to determine the type or model to be sold. At a minimum, the product to be sold should be described as specifically as possible.

Necessary or desirable location specifications may be included in this section for manufacturing, wholesale, and retail ventures. Site characteristics, such as proximity to raw material suppliers, customers, transportation, educational institutions, and other directly related resources should be specified for manufacturing and wholesale projects. For retail ventures, store location is a crucially important factor in the success of the business. The actual store location, or possible locations, will probably not be specified until the business plan is prepared. Retail store location is addressed under market analysis, because location profoundly affects the market for most stores. For an existing business, the feasibility study should review what its product and/or service has been. If any changes are to be made, these should be noted and explained.

Total Market

Having specified what products or services will be provided by the venture, the market analysis must answer the following two questions: (1) Who will buy the venture's product or service? (2) What are the characteristics of its potential customers? The answers to these questions will define the venture's primary market, those who will be its major purchasers or users, and its secondary markets, that is, its other potential customers. Trade journals and suppliers, community surveys, or surveys of local business may provide much of this information.

The geographic limitations on the venture's market must also be assessed. If it is a retail store, for example, it may be limited to a few blocks or the surrounding community. A manufacturing project may have a regional, or perhaps even a national, market. Trade associations and other knowledgeable sources can help define the market area. Also, check with business schools, libraries, and the chamber of commerce.

Once the characteristics of potential customers and the geographic limitations have been outlined, the size of the total market can be estimated by determining how many people (or firms) with those characteristics are located within the specified geographic area. Population statistics (e.g., population census) and various business censuses (e.g., census of manufacturers) should be helpful sources for these figures. This information, along with the information gathered under the competition analysis, will allow the community to determine how many buyers in the market will actually become customers (the market share). Suppliers, trade associations, the chamber of commerce, state and federal agencies, and other business owners in the area can assist in making these projections.

It is important to remember that many factors affect the size of the total market and the size of the market share that the venture can expect to capture. Although some of these are beyond the control of the community (for example, demography, economic conditions, and competition), many, such as price, product, product mix, and location, are within their control. Location has already been discussed, concerning its influence on a venture's potential market. The price of goods or services is another important element over which the community has little control. Although the price must have, as its base, the cost of providing the good or performing the service, the price is also determined by what market the seller is targeting.

The venture can be designed to target certain customer groups where the market analysis shows the greatest, or most feasible, demand. If, for example, the venture is based on expanding a nutrition program into general food services, it can investigate the size of several markets—or market segments—that exist for basically the same product. Examples would be senior citizen centers, child-care centers, and institutions such as schools, hospitals, and businesses. It can then determine whether enough demand exists in any of these segments alone, or in several combinations, to support the venture. (Remember that demand means not only need or desire for the productor service but also the ability to pay for it.) The organization can then choose which market segments to target for the highest possibility of success.

While conducting the total market analysis, the organization should be asking what choices exist and which ones will make the venture the most feasible. Once this has been determined, how much competition there is for that market should be researched.

Competition

When defining the competition, one must look not only at who sells the *same* product but also who sells substitute or alternative products. The local chamber of commerce and state and federal business censuses can help locate competitors and provide vital information about them. Ideally, the following information on each competitor should be compiled, if available:

- Location
- Products
- Customers

- Market share
- Sales and market percentage
- Competitive advantages
- Analysis of how they compete
- Growth potential

The significance of location depends on the type of business being analyzed. If the venture is a grocery store, it should probably not be located next door to its major competitor. Retail clothing stores, however, often benefit from being grouped near each other as the group can attract more business than each store alone. The state development department, the chamber of commerce, and business schools and libraries should have information on the location needs of various businesses. Be careful not to eliminate competitors outside the venture's defined geographic market area. Customers may go out of their area to shop if their alternatives are limited or if the more distant store has better products or lower prices.

Under the heading of "Products," one should try to define what similar goods and/or services competitors sell. One should determine what kinds of customers competitors attract: For example, does one child-care center in the area attract mostly parents who have the time and inclination to volunteer? Does another serve single-parent families or families in which both parents work full-time? What market segments are served by each competitor?

Competitors' market shares relate to how much, numerically, of the total market each has been able to capture, that is, the volume of the business. The market and sales percentages tell how significant the market share figures are. If a business serves 2,000 customers in a year, is this 5%, 10%, or 20% of the total market? Keep in mind that a single product or firm seldom captures all of a market.

Sales percentages can be determined by establishing how much consumers in the defined market spent on the product during a given period (total sales), and how much was spent buying the product from each competitor. The chamber of commerce, local business schools, and/or the library may have resources that specify these figures by individual firms.

Competitive advantages are the keys to a firm's success. How does each business attract its customers? Is it in a convenient location? Does it offer free parking or generous credit terms, free delivery or installation? Are its prices lower, or does its product have a marketable difference in quality?

To determine the competitive advantages of different firms, try calling directly and also watching their advertising. What image are they projecting? What services do they offer? Learning about the competition is the first step in determining how the venture can compete. In addition to the kinds of advantages included in the questions above, a venture may also be able to capitalize on its relation to the community as a whole.

Under competitive advantages, one should also analyze how competitors may respond to the venture. Are they likely to cut prices, increase advertising, or expand certain product lines? How much additional risk do competitors' possible responses impose on the venture? What competitors do will probably relate to how they currently market themselves.

A competitor's growth potential does not necessarily mean the possibility of expansion in the physical size of the store or facility, but rather its potential for increasing sales and market percentages. This can be assessed from an analysis of its competitive strategy or advantages and the projected growth of its market segments or buying power. If accessible, the growth patterns of the business over the last several years can also be analyzed.

The information above will provide a realistic assessment of the venture's total estimated market share. Once this has been determined, it is possible to project the sales volume, a critical element in its financial feasibility.

Financial Analysis

Numerous publications such as Kidder, Peabody's *Economic Development Finance* (1986) go into depth about different aspects of financial feasibility. Richard Bingham et al. (1990) in *Financing Economic Development* also offer a useful compendium specifically addressing local economic development. These are available in business libraries, from government departments that focus on small business administration, and from state banks. Accountants may also be able to provide valuable assistance in understanding and completing various financial statements. The discussion here focuses on six important financial analysis tools:

1. Estimates of initial financial requirements
2. Break-even analysis
3. Financial scenarios

4. Projected profit and loss statement
5. Cash flow projection
6. Return on investment ratio

The first five of these tools are used to determine the financial profitability and debt service capacity of the project. The last tool is a measure of the project's return-on-investment capability, a measure often compared with that of other projects in investment decision making.

All of these financial projections are very sensitive to the assumptions and projections made under the other feasibility factors. For example, if the venture requires an engineer or manager (who will probably command a high salary), that salary requirement will increase the venture's start-up and operating expenses. There may be many alternative ways of operating the business that could alter its financial standing. A business may, for example, have the choice of either manufacturing some parts itself or purchasing them. Financial feasibility could depend, to a large extent, on the alternatives selected for starting the undertaking. It may be beneficial to make several projections based on different alternatives, so that they can be compared and the most feasible option chosen.

Estimating Initial Financial Requirements

The first step in determining a venture's financing requirements is to calculate all the expenses to be incurred before the business brings in money to cover costs. This includes three categories of expenses: start-up costs, initial operating costs, and a reserve for unanticipated expenses.

Start-up expenses are the costs associated with developing and setting up the business. They include market research, technical feasibility studies, the costs of hiring employees, purchasing equipment, deposits on premises, remodeling, and initial inventory. Start-up costs also include the salary or living expenses of the people starting the business. Although some of these costs may be subsidized initially, all expenses should be carefully recorded, so that the true costs can be calculated.

Initial operating expenses are the costs of running the business until it reaches the break-even point, where income equals expenses. These expenses include both fixed costs—such as rent, insurance, and other payments that do not vary according to sales—and variable costs—expenses that do vary in relation to sales, such as advertising and wages. Exactly which costs are categorized as fixed or variable depends in part on the type of business.

A reserve for contingencies is necessary to cover unexpected contingencies and changes in the venture development plans. These costs are extremely variable, although, as a general rule, they are projected at 10% of total project cost. A worksheet that can be used to estimate the venture's financial requirements is set out in Table 11.1. It may be adapted to match the venture being started. The estimates for each expense category should be as detailed and accurate as possible, though start-up costs can be filled in immediately. In order to estimate the initial operating expenses, however, the venture's break-even point must also be calculated.

Break-Even Analysis

The break-even analysis shows the relationship between total income and expenses at varying levels of production, sales, or provision of services. It indicates at what level, given certain conditions, sales will reach the break-even point. At that point, income exactly equals expenses. Prior to this, the venture operates at a loss and must be subsidized. Subsequently, it generates a profit.

A venture could have several different break-even points depending on the conditions under which it is operated. Such conditions, for example, are the number of employees, or whether the business has its own delivery truck or hires a deliverer. Changes in these factors can cause variations in expenses and income.

For the break-even analysis, expenses are calculated as total fixed costs plus total variable costs over a certain period (e.g., 1 month):

Total costs = Total fixed costs + Total variable costs

Income is calculated as the number of units sold (either products or services) multiplied by the price of the unit:

Total income = Number of units (products or services) sold
× price of 1 unit

Break-even analysis establishes whether the level of sales income will cover expenses. Once this is known, it is possible to calculate how long it will take to achieve the necessary sales level, or whether it is possible for the venture to capture that particular amount of sales. This evaluation is made in conjunction with the information gathered in the market analysis. Based on how many

Table 11.1 Initial Financial Requirements

Expense Items	Costs
Start-up expenses	
Feasibility studies	
Legal expenses	
Salaries and wages during start-up	
Hiring employees	
Deposits on premises	
Remodeling	
Equipment and machinery	
Installation	
Initial inventories	
Advertising and promotion	
Operating expenses (until breakeven)	
Wages and salaries	
Advertising and promotion	
Administration	
Rent	
Interest payments	
Loan repayments	
Utilities	
Taxes	
Insurance	
Reserve for contingencies	
Total	

months it will take to meet the projected break-even point, the operating-expense portion of the initial financial requirements of Table 11.1 can be estimated.

Developing Financial Scenarios

Once all the project's initial financial requirements are known, the local development organization should develop a series of financial scenarios based on Figure 11.1. Each scenario should be developed using realistic projections of the costs involved. Several need to be constructed using best- to worse-case possibilities. Each component of the scenario should have a carefully developed narrative that explains how this form of funding can be obtained and the chances of obtaining it. In planning a new project, the necessary type of financing must be clearly defined. Failure to do so not only results in wasted time and effort but in unnecessary confusion and, occasionally, failure.

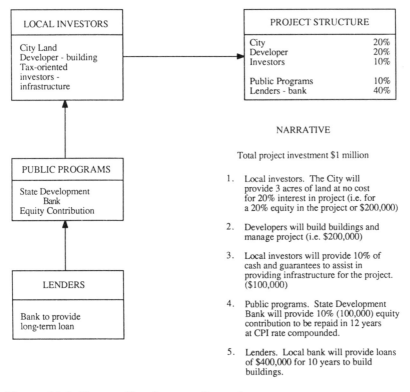

Figure 11.1. Finance Development Scenarios

From the community's point of view, a project's financing requirements are clearly defined by the purpose for which the funds will be used. The capital market institutions, however, define financing requirements by an entirely different set of criteria. To succeed in financing a project, financing needs must be defined using the criteria of both the project and the capital market.

The criteria used by lenders to analyze a financing package reflect six basic concerns:

Term. When and at what rate will the invested funds and interest be repaid? How does this correspond with the supplier's source of funds and stability?

Risk. What is the primary source of repayment and what is the probability that the invested funds will be repaid from this source? Is there a secondary source of repayment, and what is the probability that the funds will be repaid from that source if the primary source of payment fails to materialize?

Administrative cost. How much time and effort per dollar lent or invested will be required to ensure proper repayment and adequate control?

Return. What is the rate of return to the lender?

Secondary benefits. Will the investment result in the investor receiving additional deposits, new jobs, related business, goodwill in the community, or other tangible or intangible benefits?

Portfolio fit and expertise. Given the policy of the institution and its existing personnel, does the investment fit within the institution's geographic, financial, and professional area of involvement?

As can be seen, this is an entirely different set of criteria from that used in defining earlier financial requirements and will, therefore, generate very different definitions. The following example of the financing requirements for a hypothetical community business venture demonstrates the differences in definitions. A community wishes to fund the industrial estate discussed in the financial scenario. In a given project plan, it has been determined that approximately $1,000,000 of finance will be required, to be used as follows:

1. Purchase property–$200,000
2. Purchase equipment–$300,000
3. Carry receivables and inventory–$400,000
4. Early losses and contingency–$100,000
   Total–$1,000,000

This same financing package, however, when viewed by potential financiers, will look quite different, as shown below:

1. Real estate loan, 20-year, minimal risk, low administrative cost, low return, no secondary benefits, good portfolio fit–$200,000
2. Secured equipment loan, 5 years, fair risk, fair yield, low administrative cost, will receive depository business, fair portfolio fit–$100,000
3. Accounts receivable factoring, annual renewal, acceptable risk, high administrative cost, high yield, no other benefits, good portfolio fit–$100,000
4. Required equity to secure above–$200,000
5. Additional required equity–$400,000
   Total–$1,000,000

In order to successfully prepare a financially realistic project plan, financial requirements need to be marketed. To avoid wasting a tremendous amount of time and goodwill in locating the right financial sources, the project's financing must use the market's criteria.

The Profit and Loss Statement

The venture's break-even point is determined by calculating its income and expense projections. These can be laid out graphically, in a quarterly time frame, on the venture's projected profit and loss (P&L) statement, sometimes referred to as an "income and expense statement." A sample P&L statement is provided in Table 11.2.

The P&L statement is primarily a quarterly projection of operating expenses, combined with a quarterly projection of income (sales minus the cost of sales). Cost of sales includes the costs directly incurred by the business in producing the goods sold. In a manufacturing operation, this includes the cost of converting purchased materials into the finished product—raw materials, direct labor, equipment repairs, and so on. In a retail operation, costs of sales may simply include the purchase cost of the inventory that was sold. Trade journals and associations provide an average cost of sales for many industries. These can be used for comparison or rough estimation, but should be adjusted to reflect the venture's specific circumstances.

When calculating quarterly estimates for the P&L statement, remember to take into account seasonal variations in sales. A business that sells swimming pool equipment, for example, will have peak sales in summer, whereas furnace suppliers may peak during the winter months. If a firm is manufacturing a product with seasonal sales fluctuations, then an entire year must be shown.

Expenses will increase several months before the sales peak, as the venture begins to increase inventories. Seasonal variations can be estimated from trade journals and associations, and from talking with local suppliers and businesspersons. New businesses typically have a slow initial period while sales are being built up. This should also be taken into account when calculating projected sales. The months when large payments fall due will have increased expenses that should be included in the projections. The P&L statement should be accompanied by a statement of the assumptions used in making the projections. This is especially important if several

Table 11.2 Projected Profit and Loss Statement

	1st Quarter	2nd Quarter	3rd Quarter	4th Quarter
Total net sales				
Cost of sales				
GROSS PROFIT				
Variable expenses				
Salaries/wages				
Payroll taxes				
Security				
Advertising				
Dues and subscriptions				
Legal and accounting				
Office supplies				
Telephone				
Utilities				
Miscellaneous				
Total variable expenses				
Fixed expenses				
Depreciation				
Insurance				
Rent				
Taxes and licenses				
Loan payments				
Total fixed expenses				
TOTAL EXPENSES				
NET PROFIT (LOSS) (before taxes)				

variations of the venture are being tested within the financial feasibility study. It is also a good idea to develop several P&L statements for each variation on the venture: an "upside" projection that shows what it will look like if everything goes perfectly; an expected statement based on reasonable assumptions and projections; and a "downside" statement that shows what it will look like if the venture runs into a lot of problems, or if circumstances vary a great deal in unexpected, and negative, ways. Once all the categories are filled in with projected figures, the total expenses are subtracted from the gross profit (sales minus cost of sales). The resulting figure is the venture's projected monthly net profit before income taxes.

Cash Flow Projections

Cash flow projections are similar to P&L statements except that they deal with the cash a business will actually have on hand and the bills actually due each quarter rather than all of the income—some of which may be in the form of sales made on credit—and all of the expenses—some of which will not be paid during some months. The projected quarterly cash flow indicates whether the venture will have enough cash to pay its expenses each quarter, or whether it will need short-term financing during some periods. A sample cash flow worksheet is provided in Table 11.3. "Cash flow cumulative" equals the cash flow quarterly for that quarter of the column plus the cash flow quarterly from the previous quarter. As with the P&L statement, you may want to make several projections based on how different circumstances, such as leasing or purchasing equipment, will affect the venture's projected cash flow.

The Projected Return
on Investment Ratio

The return on investment (ROI) ratio estimates how much money a project will make compared to how much money was invested. It allows potential investors to compare whether the rate of return on one venture will be greater than the rate of return on the same money invested somewhere else. The ROI ratio is considered one of the best criteria of a venture's profitability and as a key measure of management efficiency. Once the venture is operational, its actual ROI can be compared with industry averages as a measure of monitoring the venture's performance.

The ROI ratio is calculated by dividing net profits (before income taxes, if any) for a certain period by net worth at the end of the period. Projected net profit can be obtained from the P&L statement. *Net worth* is defined as the owner's equity, or the owner's investment; it is simply the business's assets minus liabilities. It does not include long-term debt, although in some cases it can include subordinated debt.

ROI is expressed as a percentage. For example, if net profit before taxes for a 3-month period is $5,000 and net worth at the end of the period is $20,000, then the ROI ratio is 25%.

Table 11.3 Cash Flow Projections

Start-up Prior to Loan	*1st Quarter*	*2nd Quarter*	*3rd Quarter*	*4th Quarter*
Cash (beginning of month)				
Cash on hand				
Cash in bank				
Cash in investments				
Total Cash				
Income (during month)				
Cash sales				
Credit sales payments				
Investment income				
Loans				
Other cash income				
Total Income				
TOTAL CASH AND INCOME				
Expenses (during month)				
Inventory or new material				
Wages (including owner's)				
Taxes				
Equipment expense				
Transportation				
Loan repayment				
Other cash expenses				
TOTAL EXPENSES				
CASH FLOW (end of month)				
CASH FLOW CUMULATIVE (monthly)				

$$\frac{\$5,000}{\$20,000} = 25\%$$

In general, a ratio of between 14% and 25% is considered a desirable return for investment and possible future growth. There are other, more complicated, techniques for calculating ROI. Check with a business analyst or accountant to determine the most appropriate formula.

Cost-Benefit Analysis

Cost-benefit analysis is a complex subject beyond the scope of this book. It is best dealt with in only general terms. The following review of the basic notions of cost-benefit analysis should be supplemented by more extensive and detailed discussions. A selected bibliography that deals with the economic analysis of projects is provided at the end of this chapter.

A cost-benefit analysis examines a proposed project's net contribution to the local economy and community. It compares the local community's potential success at achieving fundamental objectives by pursuing a particular project against opportunities lost because resources were committed specifically to that project.

There are various techniques of cost-benefit analysis. The three most commonly used, however, are the net present value (npv), the cost to benefit ratio (c/b), and the internal rate of return (irr). Each of these techniques, and in fact the entire concept of cost-benefit analysis, has basic limitations. Before mentioning these limitations, however, the various techniques will be outlined briefly.

The first step in preparing the cost-benefit analysis is to estimate the project costs. These estimates should include all capital and operating costs calculated on a yearly basis for each of the years the project is expected to operate. The projected financial statement discussed in the preceding section will be a good starting point for identifying these costs. It must be appreciated, however, that these calculations reflect the project costs to its sponsor or operative entity, not the project's economic cost to the community. To determine the project's economic cost, changes must be made to the financial cost calculations. In determining economic costs only, those payments that reflect the actual use of a societal resource must be charged against the project. Payments that simply represent the transfer of control over resources from one segment of society to another should be deducted. These transfer payments include project costs such as taxes, loan repayments, and interest repayments on loans.

Sunken costs are defined as those that have been incurred on the project before the cost-benefit assessment was initiated. They should, of course, also be excluded from the cost of the project for the purpose of deciding whether to proceed with it. Only resource costs that can still be avoided are relevant. For example, the economic merit of a project designed to complete another project does not depend on the costs already incurred but only on the cost of completion.

The next step is to estimate the benefits generated by the project for each of the years it is expected to operate. These benefits will include revenues from the sale of the goods and/or services generated, and also money from the sale of any remaining equipment at the end of the project's lifetime. Contributions to society should also be added into the calculation of project benefits. To reflect these social contributions, the project's wage payments to previously unemployed workers and the price of its services must be weighed more heavily than their actual market value. This would signify in economic terms the project's contribution to the local economy and community. It must be noted that wages are counted as project costs; therefore, in charging costs to the project, the wages paid to previously unemployed people are reduced by an appropriate amount. The price of services, however, are counted as project revenues. Therefore, in assigning benefits to the project, the revenues gained from it are increased by an appropriate amount.

The process of identifying project costs and benefits is difficult. Table 11.4 provides an overview of the complexity of the issues described. The process of adjusting cost and benefit items to reflect the relevant community values and objectives is known as *shadow pricing.* How shadow prices are determined is a question best answered by an experienced cost-benefit analyst. Once the cost/revenue (benefit) items have been adjusted to reflect the relevant social or community values, one can proceed to calculate (1) the net present value (npv), (2) the benefit to cost ratio (b/c), and (3) the internal rate of return (irr).

The net present value (npv) is, quite simply, the amount remaining after subtracting the present value of all project costs from the present value of all project benefits. The key term is *present value.* This concept can be explained as follows: Because the project's operations occur in the future, it will incur costs and produce benefits at different points in time. Hence it is necessary to express future costs and benefits in terms of their present worth, that is, in a chosen base year. This can be accomplished by using a social rate of discount, which is a specified percentage that devalues future project costs and benefits.

The benefit to cost (b/c) ratio is the present worth of all project costs divided by the present worth of all project benefits. The decision rule is simply that a project contributes to society if b/c is greater than 1. The internal rate of return (irr) is the discount rate that will make the project's net present value equal to zero,

Table 11.4 Assessing Project Impacts

Direct Impacts	Possible Indirect Impacts	
Employment	**Employment**	**Income**
Construction	Multiplier effect creating other jobs	Multiplier effect
Permanent	Reduction in other jobs Education and training	
Income	**Revenues**	**Land use**
Wages and salaries of employees	Sales	New developments around project such as housing and businesses
Low-income/ disadvantaged groups	Property tax	Increased property values
Profits	Business permits and so on	Housing requirements
Land use	**Service costs**	**Environmental/ social costs**
Changes in land values	Sewer	Transportation/traffic congestion
	Water and so on	Community amenity
	Schools	Increased air pollution damage and less "room" for additional industrial development because of air-quality standards
		Community social programs

that is, the project's net present benefits equal to its net present costs over a given period of time.

There are limits to cost-benefit analysis. First, only those project impacts that can be quantified in economic terms can be easily incorporated into the analysis. Thus, social projects, which may produce benefits such as teaching a community how to organize or instilling a sense of pride and self-worth in the community, cannot be adequately assessed by using cost-benefit analysis. Second, the definition of certain benefits must be conceived very narrowly. Finally, in determining weights by which to value a project's contribution to society, or in determining the social rate of discount, there is too much room for discretion and arbitrary decisions. Given this flexibility, a cost-benefit analysis can be manipulated to show a social profit.

Put in perspective, cost-benefit analysis allows one to compare competing projects and alternative approaches to the same project. It is not, however, a precise measure of its actual value to society. Because the analysis can only consider some of the social consequences of quantitative project measurement, many important "community spirit" and attitudinal values are lost. In most instances, therefore, it is necessary for the decision maker to take the calculations of cost-benefit as an approximate measure of the project's actual worth. To improve the cost-benefit assessment, the above calculations could be supplemented with a descriptive analysis of the project's nonquantifiable costs and benefits.

Organizational Design

In making a choice of organizational form for the project, a development organization should consider carefully a number of factors:

- Economic development goals
- Amount of control the organization wants over the project
- Internal management and staff capacity
- Impact of potential liability to the local development
- Organization or government associated with the economic risk
- Image in the community
- Ability to obtain outside sources of public and private capital
- Requirements of tax, securities, and business law

Several of the most common organizational forms, including nonprofit-related in-house ventures, for-profit subsidiaries, joint ventures, and franchises, are examined below in terms of tax, business, legal, community, and organizational factors. This overview should not be considered an exhaustive list but rather a representative list of the variety of options available. The local development organization should first decide on its goals for the venture, and then work with a knowledgeable lawyer to find the most appropriate form to achieve those goals legally. Selecting the most appropriate legal structure for a project is not a one-time decision. As it grows and changes, both it and the local development organization may have different needs. The project's structure and its relationship to the local development organization can be altered to reflect the changed needs.

Using the Feasibility Study

Once feasibility studies are completed for each of the prospective projects, it is helpful to have each one reviewed by an independent expert and also by members of the local economic development advisory committee. The reviewers will provide different perspectives and valuable expertise. Their evaluations and recommendations for improvement will strengthen the feasibility study and also confirm that its findings are valid.

The reviewed feasibility studies should then be assessed by those responsible for determining whether to proceed with the venture. The evaluation of the feasibility studies should be spread over several meetings of the community or citizen group reviewing them. The first meeting could explain the function of the feasibility studies to the decision makers and also what they should be looking for in evaluating them. A second meeting could be used for the presentation and discussion of the results of the studies. A third meeting could be a community briefing session explaining the project's potential for the community and exploring other hidden costs and benefits. The last meeting would be for the final decision. Once a project has been selected, it is recommended that only one venture be started at a time. This will minimize risks to the implementing organization. The next step is the preparation of a business plan, if appropriate, or a project monitoring and evaluation program.

Business Plan Preparation

The venture's business plan will be closely related to its feasibility study, which will provide most of the data necessary to prepare the business plan. There are important differences between the two documents, however. The purpose of the feasibility study is to explore options for operating the venture, in order to assess the possible success of the activity. The business plan will lay out, in detail, how the venture will operate and the assumptions on which the operation is based. The feasibility study discusses different ways that the business might operate; the business plan describes the way that the business will operate. The business plan is a final check on the venture's feasibility. It is more detailed than the feasibility study and is often longer, especially if technical specifications are involved. The plan's length will depend on the business and the intended uses of the plan.

If it will be used primarily for board review and as an operating guide for an in-house venture with no external financing, then the plan could be fewer than 10 pages. If the venture is a separate subsidiary and/or the business plan will be used to obtain outside financing, then it will probably be considerably longer. *Don't make the plan so long and detailed that no one will read it.* One should strive for clarity, brevity, and simplicity throughout, while at the same time being careful not to sacrifice thoroughness. Most business plans do not need to be long.

The format of the business plan is another variable. Like feasibility studies, business plans have no standard method. What the plan looks like depends on the type of business and how the plan will be used. Regardless of the format chosen, most business plans should discuss at least the following points:

- The industry and the business (including general industry trends and the history of the business if applicable)
- The product and service that the business will sell
- The market (including market size, trends, competition, and market share projections)
- A marketing plan (including pricing, distribution, and promotion)
- Pre–start-up and start-up plans and schedules
- Manufacturing/operating plans
- Organization and management (including job descriptions, personnel needs, lines of authority, wages, and skills and experience of key personnel)
- Community benefits (including economic impacts, human development, and other social impacts)
- Financial plans (including initial capitalization, proposed financing, and projected financial statements such as balance sheets, profit and loss statements, and cash flow and break-even analyses)
- Control and feedback systems (including monitoring plans)
- A discussion of critical risks and assumptions associated with the business and the business plan

Most of these issues were at least touched on in the feasibility study, although they are covered in greater depth and detailed in the business plan. These basic topics can be covered in a variety of ways. What is important is that they be discussed somewhere in the plan. Seek advice from experts who can help decide what format to use and who understand the business project and the financing.

Developing Monitoring and Evaluation Programs

Project evaluation can lead to more efficient and effective projects as a result of systematic, careful analysis of project consequences and costs. It is an ongoing activity, beginning when a project is in the design or planning stage and continuing when the project is being implemented. It should also include a full review of the project once it becomes operational. In short, project evaluation begins right at the planning stage, and project evaluation procedures must be evident in the planning process itself. Indeed, the feasibility and business plan documents should include a separate section on the evaluation plan. This section should explicitly identify the following.

Criteria to Be Used in Postevaluation

If the main body of the feasibility document is done properly, these criteria will naturally emerge out of the time schedule, cost and budget limits, and productivity output targets. Key result areas, specified in measurable terms where possible, will be the criteria for postevaluation. This is as equally true of concrete technical projects as of social projects, though the latter are admittedly harder to quantify. The final benefits or effects the project intends to achieve, as defined by the project's explicit and implicit objectives, should also be included among the evaluation criteria. These benefits may include such items as these:

- A specified number of job opportunities for low-income members of the community
- Opportunities for management-skill development and job-upgrading
- Income redirection, given the flow of money in the community
- Venture self-sufficiency
- Facilitation of community development
- Infrastructure development
- Provision of quality goods and services
- Community education opportunities
- Community ownership, control, and decision-making opportunities

Evaluation Techniques

This section identifies the evaluation techniques to be used as well as the points in time when they are to be employed. In post-project evaluation, the evaluator's task is to measure if the planning stage did, in fact, reflect what actually occurred. This could mean a recomputation or remodification of costs and benefits in light of changing government priorities. The resulting figures would give implementers a more realistic perspective of the project in relation to community objectives. Thus, this section should lay out the cost-benefit premise as the basis for future evaluations to be set.

Other approaches or evaluation techniques include the use of control groups for comparative analysis, baseline measures, sampling, and various data-gathering methods such as field surveys, questionnaires, and interviews. These techniques, again, largely occur at the beginning of a project if a study or research design is planned. If this type of study is done, an experimental test group of consumers and a control group may be determined. Furthermore, baseline measures such as size and number of jobs to be created can be set. Tests or interviews conducted can reveal certain critical aspects or situations prior to the start of the project.

At the time of post-project evaluation, the comparisons made would be the experimental group before and after the project, the control group before and after the project, and an evaluation of results based on the critical aspects studied. Other techniques used would be financially oriented, such as variance analysis, where a study of the budgeted versus actual financial performance is made. If the project did not utilize a budget, however, such analysis would be very difficult, if not impossible.

Time Schedules for Conducting the Evaluation

Aside from scheduling implementation, proposals should schedule evaluation so that it is not indefinitely postponed or indicate an acceptable lapse time in instances where a project's desired results—especially social—are not evident immediately after project completion.

Budget for Postevaluation

This section should include, aside from the total amount and the program breakdowns, the promised or possible sources of funding and the schedule of budget releases.

Organization and Staff Requirements for Evaluation

This section should outline the size of the evaluation team, its qualifications, the reporting relationships, and access to project information and staff.

Those responsible for post-project evaluation must, by definition, assume a comprehensive point of view. Whereas the project's components may be worked on individually during implementation, the evaluator looks at the project as a package. Success or failure is determined by the achievement of the whole venture measured against its objectives. Thus the evaluation process must be "catholic"; it should not depend on one specialization or discipline.

Often, a team of project evaluators is composed, at a minimum, of both a technical and a project management expert. The group may consist of three to five individuals from various disciplines with recognized expertise in the line of the project. They are sometimes from universities or research agencies, as well as from various government offices. Their expertise and experience are of primary consideration in their selection, however, as the quality of the evaluation work will only be as good as the evaluators themselves.

Opinions differ on the relative merits and demerits of a preponderantly external evaluation team over an evaluation from within the project-operating agency. External teams have objectivity, possibly at the cost of lack of understanding, perspective, or access to background information both hard and soft; internal teams enjoy familiarity and access at the expense of possible subjectivity.

Another area requiring serious attention is the development of local expertise for project evaluation. It is advisable that local experts be trained in the techniques and process of project evaluation. In any case, once a project evaluation team is selected, team members must agree on the deadline for the complete post-project evaluation report.

Consistency: A Guideline
in Organization for Evaluation

Consistency here applies to format, requirements in project documentation, and procedure. The extent to which projects are documented at each different level or stage and the degree to which a uniform approach is utilized will facilitate the evaluator's task. Consistency in procedure allows for the routing of documents in such a way that the evaluating office has access to the various project files. A data bank with up-to-date records for projects and all other related materials would be extremely helpful to evaluators.

Illustration 11.1. Loan Fund Aids Rural Clinic

The Hill Country Community Clinic had a brand-new dental chair, but nowhere to put it. The only source of dental care for low-income patients in central and east Shasta County, and drawing patients from six surrounding counties, the nonprofit clinic had piled up a seven-month waiting list. To ease the crunch, the staff needed to remodel an office to create a third dental treatment room, convert a bathroom into a dental lab, and purchase new sterilization and x-ray equipment.

Hill Country Clinic lies in picturesque redwood country, but behind the beauty is a federally designated Health Professional Shortage Area. Medical practitioners are deserting much of rural California, and lack of access hits low-income and unemployed people especially hard.

Long accustomed to delivering quality medical and dental care on a shoestring, the clinic's administrator, Ray Hamby, applied for a loan of $15,000 to cover the renovations and equipment.

"But conventional lenders are scared off by nonprofit balance sheets like ours. We don't show any 'profit' because we plow it all back into services," says Hamby. "That's why organizations like ours need the Northern California Community Loan Fund."

The clinic turned to the Loan Fund as a last resort, after grantmakers and commercial lenders turned them down.

"Helping grassroots health care agencies is a priority for us. And while banks may find that such small loans aren't cost-effective, this is the kind of need we are set up to meet," says Paul Sussman, Loan Fund Executive Director.

Cash flow and funding present constant challenges in rural clinics like Hill Country. The clinic was founded by determined community activists in 1985. A $30,000 grant from the James Irvine Foundation financed construction of "a building strong enough to withstand our snow load but it was an empty shell," says Hamby.

Volunteer labor and ravioli dinner fund-raisers turned it into a medical clinic. The entire staff—physicians included—worked without salary for the first six months. In later years, counseling and dental services were added.

Impressed by the resourcefulness of the staff and community board, and with the clinic's record of "health care for people, not profit," the fund made the loan earlier this year.

[T]he Loan Fund is committed to helping Hill Country Community Clinic rebuild. Future possibilities include assistance with cash flow projections, loans to bridge the time between approval and arrival of disaster aid or grants, and financing for equipment purchases that banks may be unwilling to undertake.

SOURCE: McClain 1992. Reprinted with permission.

Illustration 11.2. Public/Private Partnership Helps Make Vermont Apple Chip Company a Reality

Vermont apple growers are celebrating the collaborative approach taken by a local economic development council and two young entrepreneurs who created Smith & Post's Crispy Apple Chips in September 1991.

"The apple chip story is one of building dynamic and innovative alliances between public and private partners to create economic growth," according to Chris Page, the executive director of the Economic Development Council of Northern Vermont (EDCNV). Created from Vermont Macintosh apples, the chips are a light new snack that debuted in regional markets last year.

The idea for the chips came from a local merchant who asked EDCNV staff members about the possibility of creating an "apple chip" product using the abundant apple orchards in the Lake Champlain valley. The EDCNV was studying ways to add processing value for locally grown commodities.

Soon after the query, the EDCNV's board of directors obtained public funding to study the feasibility of making apple chips in northern Vermont after they learned about a similar apple snack in California that was successful. They started with a small grant from the Vermont Agency of Development and Community Affairs, which enabled them to leverage additional public resources in the form of a technical assistance grant from the U.S. Economic Development Administration (EDA). The board retained a consulting firm that specialized in food products and suggested that: the resources to make a crispy apple snack existed in the region; the technologies were transferable to a small, private enterprise; and, the U.S. market had ample room for good-tasting healthy snacks.

The EDCNV's board set an agenda for action. They retained the consulting firm to create a private-sector business plan that could be used by investors to develop, finance, and operate the company. By fall 1989, the business plan was complete, and the EDCNV began to search for investors/owners.

"Most investors contacted were astonished, and somewhat skeptical of the Council's intent to relinquish its rights to the apple chip company," Page said. The Council maintained that it only wanted to be a partner in the economic development process. Finding investors proved to be difficult until Lawrence Smith and Peter Post "signed on" to pursue development of the company.

By the summer of 1991, 15 months after their introduction to the idea, Smith & Post developed a sophisticated new process capable of producing 50 pounds of apple chips per hour. They revised the financing proposal and the Bank of Vermont agreed to finance the venture. Participation by the Vermont Industrial Development Authority, a quasi-public lender, and the EDCNV completed the financing package. Soon after financing was in place, Smith & Post's Crispy Apple Chips tumbled off the production line into bright yellow, red, and green biodegradable bags.

Smith & Post acknowledged the technical contributions of the EDCNV by their offer to finance a community development foundation. The foundation will underwrite research into new market opportunities and seek local entrepreneurs to turn ideas into reality.

SOURCE: Kennedy 1992. Reprinted with permission.

Illustration 11.3. The Capital District Community Loan Fund

The Capital District Community Loan Fund, though not as large as some comparable funds, has grown steadily over its six-year lifetime, and the pace of growth has quickened. This year the Fund expected to surpass $500,000 in total loan capital. In fact it has reached a total of $750,000, including a recent capital grant of $125,000 from the Federal Home Loan Bank of New York. Lending activity has also increased. The Fund has made or committed eight loans to community groups thus far in 1992, more than in any prior year. At the same time it has achieved a new level of visibility and influence. This success is due not to any recent changes in approach, but to a patient building process involving many dedicated people and orchestrated by an exceptional Fund Manager.

Sue Kenney became Fund Manager in 1989 as the result of a personal quest that began some years ago in Florida, where she worked for eight years with "self-help housing" projects for farm workers. Though the work engaged her intensely, "it reached a point where it just wasn't complete any more. I didn't know quite what I was looking for, but I thought when it pops up I'll recognize it." From Florida she went to graduate school at Cornell, spent four months in Canada studying housing co-ops, then came to Albany, where, after an internship in the state legislature, she took a position with the State Division of the Budget.

In 1986, not long after coming to Albany, she read a newspaper article on the CDCLF. "The minute I read that article," she says, "I knew that was it." Before long, she was a member of the loan committee and board, and that fall she attended the second annual conference of the National Association of Community Development Loan Funds (which she has since served as board member and chair of the Membership Services Committee).

In 1989, CDCLF Manager Bob Andrews decided to leave the Fund to pursue a medical career. Sue applied for the job and was hired. "I was

ready," she says, "or a little bit more than ready."

She inherited a vigorous organization, but there were, as she says, "administrative things that needed to be put in place." For one thing, the Fund had sought and received a large number of relatively small loans from individuals (the Fund currently has 110 individual and 30 institutional investors). Managing so many investor loans, while also processing and monitoring the Fund's loans to community groups, was an administrative challenge. Sue's first focus was on accounting and management systems, communications with lenders, and procedures for reviewing loan applications and monitoring loans.

Meanwhile she was helping to establish the Affordable Housing Partnership (a nonprofit mortgage lender capitalized by local banks), which she has served as board chair and co-chair of the Policy Committee, and with which the CDCLF now works closely. Discussing this relationship, the Executive Director of the Partnership, Susan Cotner, says, "The loan fund's involvement with an organization often gives us a track record that we can lend against. And it's been great to be able to say to Sue Kenney, well, what's your relationship been with this group. She paves the way for a lot of these nonprofit borrowers."

To date most of the Fund's loans have been for housing projects—from single-family rehab loans to Albany Community Land Trust, to bridge loans to nonprofits developing housing with state and federal funding, to a loan for co-op shares for low-income residents of a mobile home park co-op. The Fund has also helped to finance facilities for nonprofit services such as day-care centers. "The challenge now," Sue says, "is to reach into new areas, to make loans to groups that may require more effort on our part because they're doing something that we're not used to dealing with." Exploring one such area, Sue has been meeting with other groups interested in promoting employee-owned businesses and very small enterprises within low-income communities—"to look at what's needed and come up with an idea of what each of us might be able to contribute to meet that need."

For most of her time as Manager, Sue has been the Fund's sole staff person, though she is now assisted by two part-time consultants, Sue Montgomery-Corey and Sue Spang. Further staff expansion is planned, but the Fund expects to remain a volunteer-driven organization—a quality that Sue Kenney carefully nurtures. She says, "That's when things really happen, when other people are a part of it. Albany has turned out to be a very supportive place for something like this. There are a lot of very substantial people who are there when you need them, and that makes a big difference for groups like ours."

SOURCE: White 1992. Reprinted with permission of *Community Economics*, a quarterly publication of The Institute for Community Economics.

Summary and Conclusion

The preparation of a detailed project plan is a critical juncture in project development. As the culmination of project viability analysis and preliminary design (i.e., action-plan building), the detailed project plan provides a comprehensive analysis of all aspects of the venture. It examines whether or not it is possible to implement the project (within the detailed feasibility studies), how in fact it will be implemented (within the business plan), and also how the project will be monitored to enable a comparison of its anticipated and actual outcomes.

Local governments may complete these stages or hire persons with the expertise to accomplish them. In most instances, localities will be able to review the project feasibility studies conducted and submitted to them. Few details can be overlooked in project development; if they are, the project may fail. Regardless of whether tax funds or private funds are used, project feasibility studies are important—bad investments take up space, discourage other investors, and affect the future of the community. Sound, thoroughly researched project plans should form the basis on which decisions are made. After the detailed project plans have been completed, the next step is overall development program preparation and implementation.

References and Suggested Reading

Berne, Robert, and Richard Schram. 1987. *The Financial Analysis of Governments*. Englewood Cliffs, NJ: Prentice Hall.

Bingham, Richard E., E. Hill, and S. White. 1990. *Financing Economic Development*. Newbury Park, CA: Sage.

Bruggeman, William B., and Leo Stone. 1981. *Real Estate Finance*. Homewood, IL: Irwin.

Greenwood, William, S. Haberfeld, and L. Lee. 1978. *Organizing Producer Cooperatives*. Berkeley, CA: Economic Development and Law Center.

Hill, Edward, and Nell Ann Shelley. 1990. An Overview of Economic Development Finance. In *Financing Economic Development: An Institutional Response*, edited by R. E. Bingham et al. Newbury Park: Sage.

Kennedy, Carolyn. *Economic Development Digest*. 1992. 1(1).

Kidder, Peabody and Co. 1986. *Economic Development Finance*. New York: Author.

Lafer, Stefan. 1984. *Urban Redevelopment: An Introductory Guide*. Berkeley, CA: University Extension Publications.

Lovelock, Chris, and Charles Weinberg. 1978. *Readings in Public and Non Profit Marketing*. Stanford, CA: Scientific Press.

Magee, Judith. 1978. *Down to Business: An Analysis of Small Scale Enterprise & Appropriate Technology*. Butte, MT: National Center for Appropriate Technology.

Malizia, Emile. 1985. *Local Economic Development: A Guide to Practice*. New York: Praeger.

Mancusco, Joseph R. 1983. *How to Prepare and Present a Business Plan*. New York: Prentice Hall.

Maryland Department of Economic & Community Development. 1984. *The Business Partnership in Maryland: Programs and Services for Maryland Business*. Annapolis, MD: Author.

McClain, Judy. 1992. Loan Fund Aids Rural Clinic. *Community Notes*, Fall/Winter.

Presidential Task Force. 1982. *Investing in America: Initiatives for Community Economic Development*. Washington, DC: The President's Task Force on Private Sector Initiatives.

Rados, D. 1981. *Marketing Non Profit Organizations*. Chicago: Auburn House.

Rolland, Keith. 1982. *A Survey of Church Alternative Investments*. New York: Interfaith Center for Corporate Responsibility.

Rondinelli, Dennis A., ed. 1979. *Planning Development Projects*. Stroudsburgh, PA: Dowden, Hutchinson and Ross.

Schaar, Marvin. 1980. *Cooperatives, Principles & Practices*. Madison, WI: Cooperative Extension.

Schram, Roger, ed. 1983. *Financing Community Economic Development: A Resource Economic Development Workshop*. Ithaca, NY: Cornell University.

Smith, Nathan, and M. Ainsworth. 1985. *Ideas Unlimited*. Sydney, Australia: Nelson.

White, Kirby. 1992. The Capital District Community Loan Fund. *Community Economics* 26(Fall).

Zdenek, Robert. 1983. *Resources for Community-Based Economic Development*. Washington, DC: National Congress for Community Economic Development.

12 | Institutional Approaches for Local Economic Development

    Institutional approaches or structures for local economic development have received considerable attention in recent years. One reason for this is that more local, regional, and state economic development institutions are being formed. The structure of the economic development program varies according to whether it is a state, substate regional, or local city or county-wide organization. In some instances these organizations merely coordinate activities in the public or private sector and in others the development organization is the de facto project developer. There needs to be a clear distinction between a plan for economic development and a plan for short-term measures, no matter how urgent, to meet immediate community needs. Planning for economic development intends to bring about a lasting and continuing change in the local economy, so that it will better serve social objectives. Economic development is an institution-building process. As a result, it requires the establishment of *planning systems and institutions* that can manage the development process over extended periods of time. The planning process, not the plan or document, is significant.

An institution with specific responsibility to coordinate each step of the local economic development process is essential. The development strategy plan, as well as the process itself, requires fiscal resources, technical expertise, leadership, and imagination. Some type of fully staffed, locally based institution must be available to assist in identifying and mobilizing all these resources in order to carry out strategic planning. Matt Kane and Peggy Sand summarize the essential characteristics of an economic development organization with these words:

People and organizations with vested interests in an area's economic development must be drawn into the formation and policy processes for an economic development organization. Different stakeholders will have different goals that must be made explicit and considered. The city government and private businesses certainly need to play a role. But so, too, do downtown and neighborhood representatives, labor unions, city residents, utility companies, environmentalists, other area governments, and officials from local universities. The list of stakeholders will vary from city to city. Some of these stakeholders may be important in setting general directions for an organization, whereas others may be needed for its day-to-day operations. A community's economic development organization may involve stakeholders in a formal or informal manner. But one way or another, involvement itself is important. Without it, an organization may find it lacks political support to rally the community behind its objectives and programs. (Kane and Sand 1988).

The planning model or approach is as important as the process. In this chapter, the basic organizational requirements for economic development planning are reviewed. In addition, this chapter provides case studies of various development organization types.

Organizational Requirements for Local Development

There are several basic ingredients for any local/regional economic development organization. Table 12.1 describes these components and the activities associated with each of them.

The coordinating organization for economic development is best thought of as a set of functions within a structure rather than as a single institution performing all tasks itself. (That is, the activities associated with each of the columns in Table 12.1 can be performed by staffing the organization or by the staff of another existing organization.) For example, the board of directors' function might be performed by the existing city council or a subcommittee of the city council. Alternatively, a group of local governments across a region could form a combined board.

Similarly, the executive director might be a full-time position held by a second staff member of a local government or regional organization. Of course, a full-time professional director would be best. The marketing and finance functions, however, might be provided by the local chamber of commerce or a consortium of

Table 12.1 Components of Economic Development Organization

Leadership Board Director		
Economic Analysis and Planning	*Marketing and Finance*	*Human and Community Resources*
Assessment	Promotion	Profiles
Forecasting	Project development	Education and training
Strategy	Financial planning	Community services
	Financial packaging	Regulation analysis
		Local government coordination

local businesses and financial organizations. A local university or college may provide ongoing business and economic analysis and the existing employment services agency might assist with human resource and community assessments.

Clearly, there is a wide range of effective organizational designs. The type of organizational form chosen depends on the size of the community and the level of sophistication of its institutions with respect to economic development. The specific form should be based on an analysis of the potential roles various institutions might play in the development strategy decided on by the community. Regardless of what form the development organization takes, the essential point is that it should have sufficient *authority* and *resources* to undertake at least the following activities:

- Research—to provide background information on the area's needs
- Information provision—on identified target activities
- Marketing—customized according to specific development strategies
- Coordination of the activities of other groups important to the achievement of the overall development strategy

Funds should be balanced between those devoted to staffing and those provided to staff for work-related costs. Too much funding devoted to salaries can, for example, result in staff not carrying out important tasks because they cannot provide project funding. A well-developed local coordinating body for economic and employment development will have a clearly written definition of management and staff roles and responsibilities. The essential responsibility for overseeing the planning and venture selection process in the organization must be clearly set forth. The person in charge of analysis must be responsible for preparing and monitoring reports of analysis. Any new roles and responsibilities should always be explicitly identified and added to the work load of a particular staff member.

A well-developed local economic development organization also has strong financial planning controls and a forward-looking financial planning system. The financial system must be able not only to look back and describe what happened in the past—as is necessary in grant reporting—but also to look forward and make predictions so that the organization and its ventures can accommodate and react quickly to changes. The financial system should be structured so that overhead and administrative costs are not hidden but allocated fairly and openly between the various programs. Strong financial controls enable the organization to keep accurate records of its finances and pinpoint trouble spots quickly.

The local economic development organization should also have strong economic and employment planning component. This will enable the organization to look to the future and place current and short-term community interests and needs in the context of a long-term perspective on community revitalization and independence. The planning component also allows the organization to increase systematically to the capacities of both the community and the organization for economic development activities. Effective targeted marketing programs for development activities are also crucial for success.

A well-developed organization will have a strong connections with the private sector and good relations with all the social groups within the community. This helps ensure that the organization meets the needs and expectations of these groups and enables the organization to make use of the community's resources to create healthy ventures. The support of the private sector is essential for obtaining business advice, financing, and also possible contracts. In addition, the support of the wider community can strengthen the organization's ability to obtain public and other grant funds and can demonstrate that the organization is involving the disadvantaged in the local development process. An appropriate legal structure should form the basis of a well-developed, local economic development organization. Appropriate legal structures that reflect the needs and goals of the organization or venture are as simple as possible. This means that the economic development body has no external structure for its venture but houses it internally, or that it sets up its venture as a separate subsidiary. Legal structures should allow the organization the amount of necessary control to monitor its investments effectively without interfering in their internal operations. Legal structure should follow function and, therefore, should be selected after the venture, in order to ensure that the structure is appropriate.

Leveraging the nonmonetary resources (e.g., goodwill of the community and political connections) allows the well-developed local economic development organization to make the most of its resources. Indeed, the support and cooperation of local, state, and federal officeholders is essential. The organization identifies and develops additional ventures through vertical and horizontal integration, such as starting a laundry to service an already operating nursing home. The organization may also spin off ventures, when appropriate, to free resources for the support of new ventures.

Finally, a key aspect of the well-developed, local economic development organization is its aggressive identification of new areas for growth. For continuing success, the organization must constantly move forward, identifying new ventures and activities that will help it meet its goals. Local government authorities and/or neighborhood associations must, in order to create new development initiatives, employ expert officers to act as project coordinators, economic development specialists, or persons working under similar titles. These individuals will in the following pages be referred to as *economic development specialists*.

The Economic Development Specialist

The appointment of a local economic development specialist to plan and/or direct the local development program has numerous benefits for the community, irrespective of whether a decision is made to form a new organization to guide the development process. This position helps bring both focus and commitment from the total community. The position also brings visibility for the community in wider economic and political circles. The duties of development officers vary according to the sophistication of the development system in a community. In some instances, development officers are executives who direct other specialized staff or large regional development organizations. However, most communities are likely to require an individual who not only can guide the development process but also accomplish the following tasks:

- *Research*–to assess community economic and social needs
- *Planning*–to organize people and information, and set goals and priorities
- *Management*–to develop and promote specific projects involving public-private partnerships
- *Leadership*–to facilitate the effective functioning of the local development board or commission as well as lead staff

Institutional Approaches to Local Economic Development

As mentioned previously, the actual structure of a local development organization is dependent on community circumstances. It is, however, extremely important for everyone to understand the old management principle–"form must follow function." Too frequently, communities adopt a structure based on reading about the experiences of other places. The correct procedure is to design an institutional form that fits the political and economic situation. First, however, a community must appreciate that, regardless of the form the development organization takes, it is ultimately an "enterprise" and must have sufficient capacity to perform. There are two important characteristics that development organizations must possess:

1. *Authority*–the legitimate power to act on behalf of the local government, community, unions, businesses, and other constituent groups
2. *Resources*–staff, financial access, technical assistance, information, and other resources required by local economic and employment development projects and programs

For a development organization to be successful, it must be able to use a combination of authority and resources to facilitate work. If an organization exists in name only and has only nebulous coordinating responsibilities, it is unlikely to gain the respect of the business community. The community organization has the responsibility of specifying the limits of such authority at reasonable levels. The absence of delegated responsibility will mean that the organization cannot operate effectively and that the development officer spends too much time obtaining permission to function effectively.

A Typology of Development Organizations

There are three general types of development organizations. First, there are *government agency* development organizations. These are components or complete delegate agencies of city government. Second, there are *private development associations,* which are sponsored by local/regional businesses and operate with the permission or endorsement of local government. These are private bodies usually affiliated with, or a component of, chambers of commerce, manufacturers, or other similar bodies. Finally, there are *local development corporations.* They act as semi-independent bodies that coordinate and actually manage development projects for or with local government. These organizational approaches are seldom observed in their pure form; nevertheless, they are described here as "archetypes." Each has its advantages and disadvantages. The major advantage of the government agency is its connections to the political system and, as such, it has access to the political resources of local as well as state and federal governments. Its chief disadvantage are delays of government bureaucracy. In local development planning, where speed is essential, this may be a fatal flaw unless local government genuinely supports the agency through city council actions.

Private development organizations can act quickly, but they are bound by limited interests. Generally, they are only concerned with the promotion of existing private business and with the real estate and investment opportunities associated with new firms. Private organizations seldom involve themselves in wider employment and community welfare activities. Private groups, however, can function far more effectively than government bodies in economic sectors such as tourism and retail development.

Development corporations, or "joint power" organizations, that involve government as well as business and community are the most used form of organization because such organizations continue to enjoy government and private support. In some instances, however, the organization is given too little power or responsibility from either the private sector or government agencies. When considering these forms, the following questions should be kept in mind: What role is the body to perform? What resources are available for the organization? What role is the council going to play in directing or controlling the body? What kind of staff leadership is available for the organizational form selected?

Economic Development Agencies as Units of Local Government

Some local governments consider economic development to be a regular responsibility of government that should be incorporated into its organization structure. A development department is a very comprehensive organization, and the staffing required to run such an organization is effectively out of reach for most small local government authorities. Large cities have fully staffed economic development departments. This structure has the advantage of close communications, as shown in Figure 12.1.

Independent Private Development Agencies

This type of agency has existed for some time in various local government areas. In many instances, the tourism or retailer associations have evolved from more elaborate development bodies. The chamber of commerce or some existing business group usually acts as the nucleus for this organizational framework. These organizations are effective as lobbyists for local interests with the local government. They may also raise private funds and become involved in high-risk ventures because of their status. The major benefit to the community of such a body is that it puts peer pressure on businesspeople to get things done for the community. Some smaller communities have decided that this is the best approach because the local government can augment these bodies by providing land or facilities but, at the same time, cannot become a risk to taxpayers' capital by becoming directly involved in deal making. (See Figure 12.2.)

Economic Development Corporations

The economic development corporation is an excellent vehicle and should enjoy strong support from both the public and the private sectors, although there are instances where this support is lacking. Nonetheless, this institutional approach makes it possible to bring together the complete resources of the community if the public and private sector have equal stake in the corporation.

The structure of a local development corporation can be very sophisticated or relatively simple. Many economic development corporations have moderately complicated structures. The most important feature of this institutional form is that it can perform

Figure 12.1. Model Structure of Development Organization that Is a Unit of Local Government

all of the tasks local government delegates to it while acting as a private body. For example, it can:

- administer development funds from both public and private sources
- manage industrial estates or commercial facilities for the government
- operate parking facilities and other services as joint public-private ventures
- enter into contracts and borrow funds for various development projects
- engage in marketing and promotion activities
- provide "one-stop" business service
- act as a small business assistance center
- provide marketing and technical assistance for local firms
- sponsor industrial and commercial attraction efforts

Clearly, the local development corporation is the most flexible structure for a community. Many of the area or regional development programs are of this type. (See Figure 12.3.)

Summary of Institutional Approaches

Communities, regardless of size, need to select an organizational form that meets their requirements. None of these forms is set in concrete; they can evolve over time, with components and functions being added or deleted as the need arises. There is no point in "reinventing the wheel," however. Adequate development institutional experience is available nationwide for communities to select a development framework to meet their needs. After the

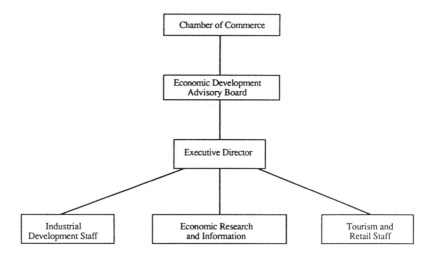

Figure 12.2. Private Development Organization Model

type of structure is selected, civic leaders can then "tailor it" to local circumstances.

Public-Private Partnerships

No matter what organizational structure is selected, public agencies and private firms have to enter into new relationships to make the development process work. This approach is much more than the public sector merely offering cooperation to the private sector to facilitate economic activities for private gain; it is far more than occasional meetings between the municipal council and local business organizations, such as the chamber of commerce. Although these activities are important, and perhaps integral to good business/government relations, they do not constitute true partnerships among the sectors. Partnerships are shared commitments to pursue common economic objectives jointly determined by public, private, and community sectors and instituted as joint actions. Analysis of successful partnership efforts suggests the following guidelines:

1. A positive civic culture that encourages citizen participation and is related to the long-term employment concerns of the community. The goals for the development process must be shared

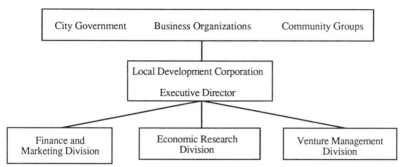

Figure 12.3. Local Development Corporation Model

among the community. Civic institutions that help create jobs and stimulate the economic base can form real partnerships, whereas self-serving business promotion or traditionally insular government—without a development role—cannot find any common ground for a partnership.

2. A realistic and commonly accepted vision of the community based on the area's strengths and weaknesses as well as on a common conception of the area's potential. This is the most important area for partnership formation. Without a common understanding of what the community has or what it can become, it is impossible to build a better community. Further, unless there is a realistic vision of the area's potential, the community will never come together to achieve its goals.

3. An effective civic organization that can blend the self-interest of members with the broader interests of the community. Enlightened self-interests are undoubtedly the sparks that light the most action. If the interests of civic leaders, both individually and collectively, can be channeled through some structure in order to achieve what is in the common interest of the total community, the development process will work.

4. A network of key groups and individuals that encourages communication among leaders and facilitates mediation of differences among competing interests. This network builds respect and confidence in the community. It allows business, labor, and government to work out their differences in private rather than in public, thereby allowing the focus of public discussion to be on the areas of agreement rather than on the problems of poor relationships.

5. The ability and desire to nurture civic entrepreneurship, that is, to encourage the risk-takers and build their confidence. Nothing is more damaging to the notion of economic development than communities that dismiss the work of effective people. It takes a few active and motivated persons and bold action to move an area or community into new job-creating ventures. If these people are not rewarded and encouraged, then the development process will stop and the community suffer the loss.

6. Continuity of policy, including the ability to adapt to changing circumstances and reduce uncertainty for business and individuals who want to take economic risks. Too frequently, government, in the absence of any consistent goals, pursues ad hoc policies disruptive to the development process. There are steps that can be taken to minimize this. First, the community should work on a set of development policies that act as a frame for its actions in the development arena. For example, a community might adopt policies that promote labor-intensive developments. Subsequent projects as well as regulations have to be examined to see if they fall within the framework. Second, local government and private enterprise, along with unions and community groups, should try to determine what kind of community they really want and build social and physical infrastructures accordingly.

These six guidelines form the basis for any organizational structure the community decides to adopt. Essentially, public and private partnerships are bridges of trust based on similar objectives but mindful of differences in roles. Achieving public/private cooperation is the first step toward engaging in actual projects. The projects will follow easily if the structure is there to facilitate the relationships.

Summary and Conclusion

Developing the correct organizational form is just as important as determining an economic development strategy. The form as well as the staffing will depend on the resources as well as the situation. Almost all economic development depends on public-private coordination and cooperation. It is essential that the pattern of these cooperative arrangements is worked out carefully so that all parties know what to expect.

References and Suggested Reading

Bradford, Calvin, et al. n.d. *Structural Disinvestment: A Problem in Search of a Policy* (mimeo). Evanston, IL: Northwestern University for Urban Affairs.

Cummings, Charles, and N. Glaser. 1983. An Examination of the Perceived Effectiveness of Community Development Corporations: A Pilot Study. *Journal of Urban Affairs* 5(4).

Daniels, Belden, N. Barbe, and B. Siegel. 1981. Experience and Potentials for Community-Based Development. In *Expanding the Opportunity to Produce.* Washington, DC: Corporation for Enterprise Development.

Daniels, Belden, and C. Tilly. 1981. Community Economic Development: Seven Guiding Principles. In *Resources,* edited by B. Daniels. Washington, DC: Congress for Community Economic Development.

Farr, Cheryl, ed. 1984. *Shaping the Local Economy: Current Perspectives on Economic Development.* Washington, DC: International City Management Association.

Gardner, Linda. 1978. *Community Economic Development Strategies: Vol. 1. Building the Base.* Berkeley, CA: National Economic Development and Law Center.

Haberfeld, S. 1981. Economic Planning in Economically Distressed Communities: The Need to Take a Partisan Perspective. *Economic Development and Law Center Report,* December.

Hein, B. 1987. *Strategic Planning for Community Economic Development.* Ames, IA: Iowa State University Extension.

Kane, Matt, and Peggy Sand. 1988. *Economic Development: What Works at the Local Level.* Washington, DC: National League of Cities.

Kotler, Milton. 1971. The Politics of Community Economic Development. *Law and Contemporary Society* 36(Winter).

Shanahan, Peter K. 1986. Economic Development at the Local Level: Public-Private Partnerships. *Australian Urban Studies* 13(4).

T.E.M. Associates, Inc. 1982. *Technical Assistance for the Revitalization of Communities.* Unpublished mimeo. Berkeley, CA: Author.

Weaver, Robert. 1991. Organizing and Staffing Economic Development Programs. In *Local Economic Development: Strategies for a Changing Economy,* edited by R. Scott Fosler. Washington, DC: ICMA.

Yin, Robert, and D. Yates. 1975. *Street Level Governments.* Lexington, MA: Lexington.

13 | The High-Technology Economic Development Strategy

It would be difficult, in fact impossible, to write a book in the 1990s on economic development and not offer a chapter on high technology. High technology has become the Holy Grail of the last two decades of the 20th century. The literature on attracting and retaining advanced or high-technology firms is growing and reveals mixed results. The precise methods, as well as the factors that lead high-technology firms to select one community over another, remain an issue of considerable debate (Blakely and Nishikawa 1992; Luger and Goldstein 1992; Saxenian 1987). Many analysts argue persuasively that for most communities the policy choice is either new high-tech jobs or no jobs at all (Blakely, Roberts, and Manidis 1987).

Despite the debates, public officials at every level are enthusiastically supporting efforts to attract, retain, or encourage new advanced/high-technology firms. In fact, nearly every urban region in the Western world has a technology park, high-tech strategy, and/or program of some type. These strategies use university resources and public facilities to create or incubate innovative firms with specialized technologies that can penetrate global markets. "The underlying message—though rarely stated," says Saxenian:

> is that once [the] prerequisites are assembled innovation and growth follow. Like a soufflé which exceeds the size of the initial ingredients, a region endowed with the proper mix of institutional and economic resources will be the lucky recipient of rapid high tech growth. (Saxenian 1989, 12)

Regions in the United States such as North Carolina, Pennsylvania, and California are all formulating recipes for specific technology

296

development. The goal today is not to attract a firm, as it was in the old "smokestack" chasing days, but to create conditions that will allow new technological firms to take root and make their corporate headquarters in the community, à la Silicon Valley, the most powerful metaphor in high technology. In this region of California, firms were not attracted but created. These firms have been the source of speculation, confusion, and a good deal of economic development "quackery." Nonetheless, it is a powerful concept that refuses to go away irrespective of regional science literature that refutes it in whole or in part (Saxenian 1989).

In fact, most Western nations have embarked on technology strategies that emphasize both specific technologies as well as specific locations. For example, France, Sweden, Germany, Italy, and Japan have embarked on ambitious regional technology development programs that pinpoint technologies as well as special human and physical infrastructure for specified localities. Even cities or groups of cities are formulating technology policies. San Jose, Emeryville-Oakland, and San Diego in California; Cambridge, Massachusetts; Ogden, Utah; Austin, Texas; and the North Carolina Triangle are among the best-known and universally acknowledged technology success stories. The Japanese Technolopolis along with Sophia Antipolis in France are the best examples of non-North American *techno-poles*. These special environments are the subject of many imitation policies aimed at stimulating new high-technology firms. In essence, regions and localities are embarking on technology-based industrial policies even when there remains considerable debate at the national level as to whether such policy initiatives are feasible. In fact, after several years of careful analysis of technology parks and similar instruments designed to promote economic development, Luger and Goldstein (1992) conclude:

> The overall policy lesson we have drawn from this analysis is that in many regions research parks by themselves will not be a wise investment. . . . Research parks will be most successful in helping to stimulate economic development in regions that already are richly endowed with resources that attract highly educated scientists and engineers. This is not to say that regions with less rich endowments cannot have a high technology future, but more basic long-term investments in improving public higher education, environmental quality, and residential opportunities will be needed first. (Luger and Goldstein 1992, 183-184)

Although regional technology policy has more adherents than national policy, it has no fewer risks, particularly in the area of selecting the correct sectors for technology investments. Obviously, trying to select or target any advanced or high-technology program is even more difficult than relying on traditional sectors of the economy. Nevertheless, anxious government officials are proposing and promoting specific technology approaches such as science parks and university incubators as the "seed corn" to grow or incubate new firms.

All of this activity seems to ignore the overwhelming literature and empirical evidence that suggests that high-technology development is nearly impossible to produce or purchase by traditional industrial incentives such as free land or low taxes, as Luger and Goldstein (1992) suggest. Rather, the evidence indicates that advanced technology firms are induced by the region's character. That is, the setting itself and neither the amount of money spent nor other blandishments will stimulate the creation of new firms or the attraction of existing technology-oriented enterprises. In a sense, "some places have it and others don't." The question is: Can a community that does not have the correct environment now create or stimulate it? In this chapter, I cannot answer that question categorically, but I can suggest the capacity requirements that communities need at the various stages of technology development to examine their *technology development capacity* to start, retain, and sustain a high-technology-firm environment.

The Technology Development Process

Technology development presents no instant success formula. What is applicable or successful in one region may not, if duplicated, be successful in another area. Each region must find its own place or niche in the technology environment. In fact, high tech may not be the best option for a region to adopt if it lacks the factors that can or will induce "knowledge-intensive" firms.

The range of support systems needed to encourage and foster advanced technology development are many and complex. Only a broad list of essential support structures is addressed in this chapter. The basic orientation I take is that the "soft infrastructure" (living amenities, cultural and educational institutions) leads to the hard (money and physical) infrastructure, and not the reverse. To plan for advanced technology, one must first understand how the technology process works. There are four stages, and the failure to provide the necessary supports at any one stage may ultimately inhibit the development of new products and industries. The interrelationships between the soft and hard infrastructures are depicted in Figure 13.1. In this figure a set of related social and technical forces forms the basis for a region or area's technology choice. These social, economic, and physical factors shape the technology strategy.

High-Technology Regional Choice Model

The technology process is cyclical and interactive. Each stage in the process may revert back or interact with a previous stage. For example, product failure spurs new research or a new phase of innovation. The process has its own support system requirements. In too many instances, the regional policy or program fails to recognize these distinct stages and consequently fails to create the right environment at the right time. The basic infrastructure components for a technology-based economic development strategy can be described as follows:

1. Research Base. Research identifies the basic scientific principles on which new technology industries or products are based. Research may be applied (designed to solve specific problems or have an application for a field of technology) or generic (nonspecific research in areas of continuing pertinence for a business enterprise). Targeted research may be market-driven, guided by the knowledge of market requirements; or technology-driven, guided by perceived capabilities of specific technology. Each field of research will also have its own peculiar support system. Most generic research is performed at national universities or laboratories, mainly with government funding. Most applied research is performed at large industrial labs. Many university communities have a set of research institutions with both the generic and applied/firm based institutions. Recent analysis of technology areas indicates that the communications or networks among research firms are the real incubators in economic development (Blakely and Nishikawa 1992).

2. Invention. Inventions are products that may or may not arise directly from research. Often they result from ideas that help to improve an operation's efficiency. In many cases, gadgets, devices, or programs are the result of invention. These are eventually developed and refined as new technology products. A technology milieu creates an environment that stimulates more invention.

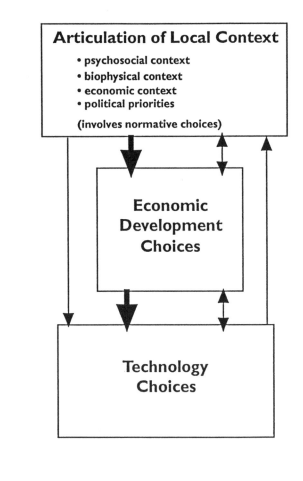

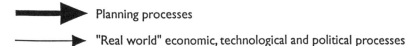

Figure 13.1. Economic Development Planning and Technology Choice

Moreover, it is important to have in place organizations to assist in patent handling as well as creative financing in the community. It is no accident that places such as Silicon Valley near San Francisco develop new firms. The San Francisco area has a very well-developed venture capital base and excellent legal services for scientists.

3. Adaptation and Development. Adaptations resulting from research or invention modify or improve existing technology. Modifications of software, for example, improve the operation's efficiency.

The second stage in the process of technological development is product development. At this point, ideas, research principles, inventions and adaptations are translated into a practical product. Prototype tests or models are built to evaluate the performance, application, and feasibility of developing the product to a commercial level. More detailed feasibility studies and some market research will be undertaken. Much prototype development work on new products and technologies is undertaken by small companies. These are often more productive and efficient than large corporations, as most of their resources are devoted to research and development. They must also survive on what they produce.

Communities and regions can assist in the adaptation and development process by providing actual incubator facilities and creating technology or manufacturing zones so that firms can find suitable space for their initial operations.

4. Innovation and Dissemination. The third stage in the process involves the commercialization of a new product or process. A company will decide to develop the product to manufacture or sell it. If it decides to manufacture, the firm may enter into joint venture or other production arrangements. Marketing, venture capital finance, and entrepreneurs will be important in the innovation and dissemination stage.

These first three stages in the technology process all represent investment costs. Through the manufacture or sale of new technology a company produces a return and recoups the cost of the product's creation. Production is the final stage of the process. Larger companies that have purchased or developed a high-tech product will manufacture and distribute it. Other companies will disseminate the production of components to subsidiaries to produce, for instance, chips and circuits. Constant marketing and monitoring of the product is essential during the production stage. Regular feedback to research and development and marketing is essential in order to continue to improve a product or terminate the production before it reaches the end of the technology life cycle.

In order to facilitate the development process a community has to provide "places to meet." These places range from convention centers to restaurants and theater areas. This "soft infrastructure" can be even more important than physical roads, sewers, and special industrial parks.

Regional/Community Support Systems for Advanced Technology

During each stage of the technological process, a complex range of support systems is required to ensure that the application of research or ideas is carried through to the production stage. The support systems required for advanced technology development are numerous, and the extent to which they are needed within the process of technological development varies. But unless the right supports are available at every stage of the process, the process breaks down and the prospect for promoting advanced technology industrial development is considerably weakened. It is important, therefore, to understand how the process operates so that the right support systems can be put in place for the different stages of technological development.

Advanced technology development requires timely and totally integrated support systems. High-technology firms have a complex and interrelated set of special requirements based on their stage of development and including their size, technology base, and global market position.

These support systems are depicted against the development stages discussed earlier in Figure 13.2. The list is based on a compilation of the latest literature on high technology and location, as well as my own field observations in implementing technology development strategies.

The Soft Infrastructure

Because information systems, education and research, and business and environmental supports form the soft infrastructure for development, these are important in starting the research stage of the technology process.

Technology is driven by information, the essential link between each stage. It would be impossible to proceed to the next stage without information, since it provides the clues about the direction in which to move. Without research, no matter how simple, a product is unlikely to be developed; without marketing information, new products run a high risk of failure; without social interaction, the flow of information about research and new ideas between scientists slows, and so does the rate of change in technology.

The information required at each stage in the technology process is different. At the research stage, buying or acquiring information is

Requirement	Research	Development	Diffusion	Production
Information Communication Strategic relationship	Social interaction Library Database	Center/hotel	Innovation centers Information bureaus	Publication Media Marketing
Human resources	Research scientists Inventors	Skilled technical staff	Entrepreneurs Venture capitalists	Management Skilled workforce
Education and research facilities	University research institute	Research labs	Conference facilities	Community college Technical institutes
Environmental quality	Quality living and working environment	Quality living and working environment	Quality living and working environment	Quality living and working environment
Government support	Government contracts for applied research Offsets	Joint ventures	Information support services	Offsets Export incentives
Finance	Research grants	Private investments and venture capital	Private investments and venture capital	Institution finance
Technology image	Quality residential areas	Reputation among luminaries	Clear identification with specialized tech firms	Top quality companies
Enterprise facilities	Research institutes University and national labs	Incubator innovation Building centers Labs	Incubator innovation Building centers Labs	Technology, business, and quality industrial parks
Infrastructure	Basic infrastructure for research facilities	Research stations Testing areas	Research stations Testing areas	Fiberoptics Airports Waste disposal

Figure 13.2. Requirements for Industries by State of Technology Process

important. At the production phase, selling information will be essential to commercial success. Information is transferred on a formal and informal basis, and special attention needs to be given to the latter. Much of this information is not technical but commercial production and marketing information.

Conference centers—some associated with hotels, others with university and research institutes—play an important part in the information transfer support system for advanced technology. Information exchange is also fostered by greater access to information networks, through libraries, information exchange bureaus, and the media. The ability to tap into networks is very important, and the community must make investments in communications technologies to provide the necessary infrastructure to help establish new and improved networks as the base for firm development.

Inducing the High-Tech Base

Creating an environment that maximizes information exchange, face-to-face communications, and the formation of strategic relationships is the key to successful advanced-technology development. These networks or channels of both information and exchange of capital and expertise form the nucleus for the other physical, social, economic, and environmental systems needed to support advanced-technology industrial development.

Human Resources

Human resources, in terms of skills, know-how, and entrepreneurship, are an important support system for advanced-technology development. The research phase of development requires a highly skilled and flexible workforce. Most people involved in research will have master's or higher degrees, and some will have multidisciplinary skills.

Social skills in dealing effectively with people from a variety of cultures and foreign languages are essential in globally based enterprises. These skills develop as the result of peer group association and interaction during forums of scientific and social interest. One of the successes of Silicon Valley is the high level of social interaction that develops between scientists in one locality. Scientists in social vacuums are deprived of innovative thought that results from social interaction. They rely on publications that are limited in what they can convey. Once a research discovery or invention proceeds to the development stage, a pool of technicians must undertake the rigorous testing and quality control for new products. These personnel do not require highly specialized skills, but they must be able to possess a careful eye for detail.

During the development and dissemination phase, the most important resource is the marketing person. Social entrepreneurial skills are an essential link between product development and production capital. Trained financial marketing depends on a human resource support system that can promote the products and bring in the venture capital. Special marketing, finance, and management skills are required in the final stage of the technology process, when a product is manufactured.

Employment mobility is an important feature of advanced technology areas. This is partly because of the rapidly changing nature and high failure rates of technology enterprises. In addition, many individuals do not want to be bound to companies and prefer contractual or project-based employment. Most businesses look on this favorably, as it reduces problems with staff retrenchments in economic downturns or at the end of technology life cycles. This mobility also encourages considerable interchanges of staff, and hence ideas, within companies and research centers, providing a convenient means for refresher courses. In most of Europe, mobility is generally not encouraged, partly because of the lack of transferable superannuation and other benefits as well as other structural rigidities in the labor market system.

Education and Research Facilities

Another important support system for advanced technology development is the presence of a university or technical institute of international standing that encompasses science and research. A university provides a catalyst for future scientists and researchers to develop their ideas and thinking in a wide field of science. Evidence from studies carried out in the United States and elsewhere suggests that although universities do play a direct role in supporting high-tech research, they have a more vital role in providing a pool of fresh minds and ideas necessary for high-tech industries.

University and Research Links

Perhaps the most important support system for advanced technology development is the establishment of research institutes. These are specialized institutes that emphasize applied research. Stanford Research Institute, for example, was one of the important ingredients in propelling the high-tech development in Silicon Valley, California. Such institutes might be privately or publicly funded and may consist of one or more centers.

Equally as important as these institutes are research laboratories and incubator establishments, some tied to major companies such as Bell Laboratories and the Wang Research Institute. There is a tendency for major companies to form, or parcel out work to, subsidiary companies whose activities specifically involve research and development. Government research centers are also important in this process. Many are involved with generic or pure research.

During the development stage of a technology product, there is a need for a pool of technical staff whose training, although not to a university level, must be sufficient to carry out skilled work. Research or technical institutes are often responsible for training skilled technical staff. Colleges with specialized courses related to science may also be needed to provide lower-level support staff for research institutes or research laboratories. Technically oriented colleges form an important role in training staff for mid- and low-tech industries.

Although universities, research institutes, and laboratories perform an important educational function in the process of technology development, strong support is also required at the secondary education level. Science students at high schools must be computer literate and have some knowledge and training in research methods to gain employment in technology-based enterprises. This support goes as far as providing computer facilities in public libraries for pupils of families who cannot afford home computers.

Environmental Quality

The quality of the environment in which people live and work (lifestyle) can have a significant effect on whether an individual chooses to live in one locality or another. Experience throughout the world indicates that personnel involved in research and development choose to live in high-quality residential environments with good amenities, and cultural, commercial, and educational facilities. Rural and small town areas, not too distant from major cities, are especially good locations.

High salaries are unlikely to attract personnel to areas where the environmental quality is perceived to be poor. Most scientists value the association they have with friends and colleagues in an environment that has a quality setting and cultural and intellectual stimuli. Similarly, executives and managers require good living environments as part of their work environment. The total environment, including recreational areas, homes, and social gathering places, is an important part of a knowledge-sharing business.

Business and Support Institutions

Although universities and research institutes play a significant role in promoting new advanced technology-based enterprises, the literature and observations indicate that major corporations and institutions that have research and development sections attached to their operations form a major stimulus to the growth and clustering of technology industries. Many large American companies have defense contracts that have been the principal stimulus behind numerous high-technology developments. Other companies in the pharmaceutical, biotechnology, medical, and chemical fields offer significant spin-offs for advanced technology enterprises.

Many state and local governments make the mistake of recruiting technology branch plants as the basis for forming a high-tech economic base. They mistakenly believe that new firms will emerge from these branch operations and headquarter in their locality. Although this is possible, the current research indicates that there is very little spin-off of new firms away from the research headquarters (with the exception of software development). If a company moves its headquarters or a major division and product lines to a community, it usually continues to use its own internal support systems for servicing the branch operations. However, a community that embarks on a strategy of incubating or starting small tech firms can create an agglomeration of technology industries within the region or locality. When this occurs, the multiplier effect of technology development is significant, because R&D businesses pick up contracts from these larger companies. Small high-tech businesses are dependent on large businesses to fund applied research and develop high-tech products. Evidence shows there is a very strong vertical relationship between small high-tech businesses and major companies. Many small start-up firms supply components to larger companies. These small companies develop a niche that remains as long as the parent manufacturing company continues to use its products. Failures in small-scale high-tech enterprises are due to major companies terminating their contracts with servicing companies when the primary product is no longer marketable. Small companies are part of the technology life cycle of product development. Many never survive to grow into large firms but become part of larger operations. There is a trend in recent years to establish innovation centers to encourage the development of small-scale businesses involved in high tech. Many of these have failed partly because they were too far removed from the major businesses they served. Innovation centers are more likely to succeed when located close to major users or developers of high-technology products and research institutions.

The importance of institutions as catalysts for advanced technology development has often been overlooked in the belief that the greatest spin-offs come from corporations. Medicine is one of the sciences in which research is expanding rapidly; hospitals are a base for much of that research. As hospitals become more specialized, there is a greater need for more precise instruments, drugs, vaccines, and equipment. Advanced technology firms have considerable opportunities to take advantage of these institutions' needs.

Government institutions that can act as catalysts for high tech are telecommunications, government science laboratories, agriculture and marine research centers, and administrative services. The last group is becoming increasingly dependent on substantial data storage and retrieval systems and processing. The obvious benefits of defense institutions are well documented, but the opportunities to develop advanced technology industries that serve the special requirements of other government institutions are numerous.

Government Support

Government's primary role in promoting advanced technology development should be supportive and not directive. Experience in several nations has shown that direct government involvement results at best in mediocrity and more often in disappointment. There is a tendency for government to try to dictate the path of research and the outcomes rather than allowing scientists to manage the process.

The role of government is to provide services, because both businesses and research personnel are attracted to world-class facilities. These scientists in turn strengthen a region's image, attracting other industries and scientists. Those areas with well-established advanced technology enterprises have world-recognized universities, well-known company names, and lists of Nobel prize winners. The first step in the long-term task of creating a technological image is the naming of facilities such as education and research institutes. The Massachusetts Institute of Technology, for example, conveys such an image. The second step is ensuring that future development is of the highest standards and environmental quality.

Government Regulations

Regulations have a strong impact on the decision by businesses to locate in one region or another. There are no rigid guidelines that can be laid down to determine the percentage of a business that can be used for one land use activity or another. Advanced-technology industries are in a constant state of change, as is the land or floor space they use. The development control system adopted in advanced technology industries should be very flexible. Planning guidelines should stress performance standards and the need for high environmental quality through landscaping, urban design, architecture, and supporting social and recreation facilities. Planning policies should give incentives to industry and not act as a total controlling mechanism. There are always other regions that will bend rules to entice industry.

Enterprise or Incubator Facilities

Enterprise facilities describe the buildings, areas, or estates necessary to carry out business associated with high technology. Some of these would be developed as part of other support systems, for example, universities, research institutes, research laboratories, and innovation and conference centers.

Depending on the type of facilities required, directors of advanced-technology companies will select a locality for either the corporate headquarters, satellite or branch offices, research and development facilities, subsidiary manufacturing center, or field offices. Some companies may combine all of the above in one location. For some enterprises, a rural location will be selected; for others, an urban, commercial, or industrial area will better suit their needs.

In a study by Saxenian (1989), in which she compared the location of high-tech industries in the United States and Britain, high-tech industries were shown to fall into two locations: those industries centered in older well-established industrial regions, and those in more open and spacious rural settings. Most new businesses, however, will follow the trend of becoming established in one of five specialized development areas. These are research parks, science parks, technology parks, business parks, and high-quality industrial estates. These are defined as follows:

Research park. A high-quality, low-density physical development in a parklike setting, where there is significant interaction between academics, researchers involved in research and product development, and commercial organizations and entrepreneurs.

Science park. An innovation center containing research and development enterprises that can also include light industrial production relating to scientific research and appropriate ancillary services.

Technology park. A collection of advanced-technology industries concerned with both research and manufacturing, located in attractive, well-landscaped surroundings and situated within a reasonable catchment area of a scientific university or major institute.

Business park. A prestigious environment, suitable for a wide range of activities, including manufacturing, assembly, sales and other office-based activities. There is no requirement for these parks to be close to academic institutions.

High-quality industrial estate. These estates have attractive features for certain forms of contemporary light industries.

There are many benefits to be gained by new industries being located in specialized, well-marked facilities, namely good services, shared facilities, high-quality environment, and flexible planning conditions. These facilities are now becoming firmly established with the image of high technology and are therefore an important support system.

Community Facilities and Support

Few scientists will choose to live in areas with a poor range of community facilities and services. Good shopping facilities, entertainment, restaurant dining, recreation, and medical services are important to their way of life. Such facilities are also important in developing associations between people involved in research and development and high-tech industries. These associations allow ideas to be expressed on an informal basis and help circulate knowledge.

As many people involved in the fields of science, research, and development are generally highly paid, most of them expect higher levels of service and are prepared to pay for them. There is an exclusiveness about high-tech personnel and where they live, recreate, and do business. The quality and range of services expected by them is therefore very important to their lifestyle.

Summary and Conclusion

High technology is the new elixir of community economic development. If you cannot find a firm, grow one. Every community in the nation is convinced that it can or should become the home to a new technology. Unfortunately, few communities can ever aspire to a high-tech future. However, almost every community can prepare itself for future technologies by preparing its human resources better and incorporating new technologies such as advanced telecommunications into its public infrastructure. The technology choices for most communities are narrow but that does not mean that they should not be examined.

References and Suggested Reading

Blakely, E. J., and N. Nishikawa. 1992. Inducing High Technology Firms: State Economic Development Strategies for Biotechnology. *Economic Development Quarterly* 6(3).

Blakely, E. J., B. H. Roberts, and P. Manidis. 1987. Inducing High Tech: Principles of Designing Support Systems for the Formation and Attraction of Advanced Technology Firms. *International Journal of Technology Management* 2(3/4).

Blakely, E., and K. Willoughby. 1990. Is it Mainly Transfer or Generation in Biotech? *Journal of Technology Transfer* 15(4).

Case, John. 1992. *From the Ground Up: The Resurgence of American Entrepreneurship.* New York: Simon & Schuster.

Galbraith, G. 1985. High Technology Location and Development: The Case of Orange County. *California Management Review* 28(1).

Luger, M. I., and H. A. Goldstein. 1992. *Technology in the Garden.* Chapel Hill: University of North Carolina Press.

Rycroft, R. W., and D. Kash. 1992. Technology Policy Requires Picking Winners. *Economic Development Quarterly* 6(3).

Saxenian, A. 1989. *The Cheshire Cat's Grin: Innovation, Regional Development, and the Cambridge Case.* Working Paper No. 497. Institute of Urban and Regional Development. Berkeley, CA: University of California Press.

Schmidt, J., and R. Wilson. 1988. State Science and Technology Policies. *Economic Development Quarterly* 2(2).

Willoughby, Kelvin. 1990. *Technology Choice.* Boulder, CO: Westview Press.

14 | The Local Economic Development Profession and Professionals

The individuals who work as economic development specialists are forming a new profession, based in several other occupations ranging from planning and economics to social work. The forerunners of this profession were industrial developers, promoters, and community boosters from the chamber of commerce or other civic organizations. Although this field has no real origins in the classical professional sense, it owes a great deal to the existing applied disciplines of geography, business administration, public finance, political economics, and urban planning. In essence, the local economic development orientation, or set of practices, is a hybrid of existing concepts, disciplines, and areas of practice molded together to form a new area of "professionalism." This new professionalism is a mixture of rational planning and salesmanship. Levy summarizes this notion:

> The academicians who wish to increase the usefulness of [their] contribution to the practice of economic development at the local level or make useful suggestions regarding state and national policy direction, must recognize the dominance of the sales side of the process in the practitioner's work. Efforts to tie local economic development into the broader context of community planning must be fitted into a setting in which sales is likely to remain the dominant mode. (Levy 1990, 158)

It is the practice, not theory, of the planning process that defines the roles and limits of this emerging professional economic developer, rather than any professional establishment. In many respects, flexibility is an asset to this field because practitioners are not restrained by rigid code. The flip side, however, is that the absence of standards makes it difficult to exclude any activities, even the most patently anticommunity activities, from being labeled as local economic development initiatives. Moreover, any government or local community group can (and sometimes does) merely change the name of their activity or the name on the door in order to label their activities as economic development.

In Search of a Professional Identity

The central components of this emerging profession are derived from the functions that the professionals perform rather than from the various disciplines from which it is derived. The economic development profession can be viewed as comprising the interrelationships of five elements: locational factors (L); organizational role (O); task functions (T); nature of the clients served (C); and the individual practitioner orientation (I). The economic development (ED) specialist role may be seen as a function of all these factors. This might be shown as:

$$\text{ED spclst role} = f(L_a, b, c \ldots O_d, e, f \ldots T_f, g, h \ldots C_i, K, l \ldots I_s, t, u \ldots)$$

A more cynical formula put forth by one economic development specialist is this:

> There is a phrase that many people won't admit to but "shoot anything that flies; claim anything that falls" [laughter]. I think a lot of people busy themselves with these retention surveys, whatever, and going to trade shows. . . . God, he is out there marketing. (Rubin 1988, 244)

Each of the five factors mentioned above has all of the practical marks described by the "shoot anything" point of view. Factors such as organization, clients, task, and individual orientations are more heavily weighted according to what the community or the specialist is "shooting at." In some instances, an individual ED specialist can achieve things the environment does not dictate and he/she can claim credit for whatever occurs. In essence, the personal variable, both skills and personality, can and will dominate practice. However, the tasks themselves may provide a productive or counterproductive environment. Luke and colleagues put it well:

The experience and working knowledge of the seasoned economic development manager are increasingly ineffective and, in many cases, even detrimental when applied to the new interconnected economic context. Competing with other cities for scarce industrial prospects creates adversarial, competitive relationships that can actually hinder future economic development. New collaborative strategies are required for several reasons. One is the expanding and crowded economic development arena . . . [p]olicy-making responsibility is dispersed and shared by a multiplicity of elected and appointed public officials. Another is a significantly reduced capacity for any one government agency or individual manager to effectively act unilaterally. A third is the slowness of policy formulation and implementation. . . . A fourth is the inevitable increase in vulnerability and openness to outside economic forces, with cities and states increasingly influenced by corporate investment decisions made in other cities on the globe. Economic development is now set in an intergovernmental and intersectoral web of pulling and pushing, and governments can seldom deal with economic problems independently. (Luke, Veattriss, Reed, and Reed 1988, 227)

Therefore, in searching for an appropriate definition/role, I will examine each of these factors in order to develop a framework for both the profession as well as professional roles.

The Community

Almost all communities or regions present different circumstances irrespective of the causes of their economic ills. Many intervening factors can influence the situation. For example, one community may be easily mobilized because there is a clear pattern of leadership while another has virtually no identifiable leaders. The goals in both circumstances may be identical and the tools or methods used by the practitioner may bear superficial similarity, but the precise mode of operation will vary based on the conditions. As a result, not only the objective circumstances but also the milieux for development are significant issues for the ED specialist.

The total community or area circumstances must be taken into consideration when assessing economic development (ED) needs. The atmosphere in which the change is to be made is as important as the change itself. As a result, the ED specialist must reach well beyond technical know-how, to help the community see itself as a social and physical entity that in many instances goes well beyond the parochial boundaries of the municipality. This identification task is a necessary and important ingredient in the development process, and failure to come to grips with the real "locational" issues can doom even the most dedicated ED specialist.

In addition to getting the economic geography right, the ED specialist must also assist the community in getting the economic causal problem right. This is not an easy task, as stated in earlier chapters. Many communities identify their problems as external, such as competing lower-wage places of foreign nations or unfair international competition. This may or may not be the case. In most instances, however, dwelling on the external will not create any new options. Therefore, the ED specialist's job is to identify the economic problems that can be solved within the context of that locality. This frequently means helping the community recognize that remedies selected elsewhere (high tech, tourism, factory attraction, and so on) may not suit the locality. In essence, the situation determines both the means of and the limits on local economic development. The ED specialist must be aware of this and use appropriate skills to assist the community to find the correct path to a sustainable economy.

The Organization

As previously discussed, local economic development activities are conducted by a variety of institutions at both the local and state levels. Primarily, the ED specialist works in municipal or multijurisdictional local government organizations. Therefore, most of the comments in this section concern practice at the local level.

Each economic development organization is formed to fulfill some preconceived mission. That mission may be clear or fuzzy, but it forms the justification for the organization's existence and imposes a limitation on the authority of the ED specialist. In many, perhaps most, circumstances, economic development organizations are coordinating bodies that take few direct actions themselves. In such instances, the ED specialist is more of a resource person than an expert. Because other organizations are carrying out the action, the ED specialist must possess the skill of encouraging without displacing and monitoring without ruling. It is a difficult role to play.

In other circumstances, the organization may view itself as a developer or development partner. In these circumstances, the ED specialist must be creative and aggressive and seek opportunities to participate in new economic activities on a broad front, ranging from housing to industrial development. The ED specialist might even be required to design new financial instruments or help firms organize themselves to take advantage of government or other development programs.

Increasingly, the ED specialist is part of a larger government or nonprofit organization bureaucracy. As a government agent, the ED specialist may have limitations imposed from a variety of sources ranging from advisory boards to elected officials. The tensions created in this environment can be overwhelming, especially when different constituencies have widely varying concepts of the role and responsibilities of the ED specialist.

Economic development organizations grow and specialize. As a result, the ED specialist may operate within a large economic or planning bureaucracy. Organizations of this type expect the ED specialist to have certain definable subareas of special skill such as housing, finance, and small business.

Finally, economic development organizations operate in a wider framework of organizations and institutions at the local, state, and national levels and, more recently, at the international level. The interorganizational scope of local economic development is expanding. The need to create links with larger or more specialized organizations such as airport authorities or international development agencies, as well as the substantial lobbying required with state and federal legislatures, provides an exceptionally dynamic situation for the ED specialist.

The Task Functions

The specialist's task is complex, to say the least, but falls into the general categories of sales and analytical methods. These activities are difficult to separate because they are intertwined in carrying out the activities of any specialist. The sales dimension involves activities associated with organizing people, selling ideas, and mobilizing resources for economic development. No amount of technically correct activity will help a community revitalize or direct its economy unless there is sufficient internal capacity. It is the process side of the ledger that helps build capacity in terms of leadership and organizational strength.

The methods dimension refers to the set of strategies and approaches that the ED specialist either fashions or identifies as suitable for the situation. These methods have been discussed in previous chapters. The methods are administered through (1) identifying community problems, (2) providing technical and analytical assistance, and (3) determining the resource mix required to meet the economic development needs of the situation, and through the process skills of (4) developing networks among individuals and institutions and (5) stimulating interaction among diverse groups to achieve a common objective.

The specific tasks of the ED specialist are as follows:

1. Building development organizations. One of the most important functions of an ED specialist is to develop a strong, viable, and continuing organization. Capacity building relates both to helping the organization gain expertise and to identifying and developing future leadership. In addition, the ED specialist must assist the organization in strengthening its network with institutions at the local, state, national, and international levels.

2. Inventory of area resources. Keeping track of community or area resources goes well beyond physical assets inventories. An area's development resources extend to its culture, its leadership, and the quality of its community social life. The ED specialist must not only know the resource base but find new ways to use it to achieve community objectives. Of course, the ED specialist must also find ways to build resources where there are deficiencies or to transform other resources to meet the need.

3. Selecting strategies. The community selects strategies, but the ED specialist helps guide the community in the process. This is one of the most important process skills of a specialist. The ED specialist must be careful to "assist" rather than push or sell a particular strategy. Moreover, the ED specialist must help the decision makers see the need for comprehensive approaches, incorporating several methods over single-component strategies such as tourism.

4. Marketing the area. Marketing a community is not like marketing a product. Products remain constant in terms of their performance. Communities change. They both add and lose capacity depending on events. No matter what role the ED specialist plays, marketing will be a component. The marketing of the place is also associated with the marketing of many other factors, for example, human resources, organizational capacity, and community incentive programs.

5. Data development and analysis. Economic development data are not always clear or clean. The ED specialist constantly attempts to find good information on the local situation, ranging from demographic to institutional analysis. Usually large-scale data sources are inadequate for the purpose. Therefore, the ED specialist must refine data from the census and other sources, conduct surveys, or identify unobtrusive methods to measure any particular development dimension. Having data and determining what they mean are also not easy tasks. As a result, the ED specialist must frequently develop ways to display complex information for the layperson. This requires exceptionally good analytical and presentation skills.

All of the above tasks relate to the institutional and locational circumstances of the community. All five are required at different times. Timing is thus an extremely important skill, and mastering good timing is a fundamental task of a good ED specialist.

The Clients

The ED specialist must work with a wide variety of clients. Clients are both individual and collective. In many respects, the total population of the community or area forms the client base. The ED specialist usually has a direct contact with a regular group and more limited contact with a wider constituent base. Generally, the ED specialist will work with a single advisory body or group and maintain contact with others through organizations or other collective agencies.

Lay leaders are usually the clientele for local economic development. Such persons are generally volunteers working for the good of the community. As laypersons, they may have little or no preparation in the field. They may well represent community groups with considerable influence but without any special expertise. As a result, the ED specialist must respect these individuals for the knowledge they bring from their constituents and use this information in program development. However, the professional must provide training and technical assistance to the group. This is a delicate role. The ED specialist needs to balance the group's lay knowledge with professional expertise.

Local officials are frequently the employers of ED specialists. Economic development departments or agencies are important components of local governments. ED specialists may direct such departments and, doing so, report directly to local elected officials.

The ED specialist's success in such circumstances depends on supporting these officials in meeting the needs of the community. This role may require anything from general economic development planning to community education to consultation with community groups.

Civic and economic development organizations are frequently the prime sponsors of economic development. These organizations, described earlier, are public-private institutions that take on the mantle of local development generally with official sanction from the local or multijurisdictional authorities. Although these groups are also organized by volunteer laypersons, they frequently include individuals with considerable business expertise and resources. The ED specialist usually facilitates the activities of this type of agency. Because such groups tend to have substantial resources, the professional role requires coordinating economic development activities, including the hiring of additional specialized expertise. In addition, the ED specialist acts as the principal contact person with other organizations and agencies.

In sum, the clientele for local economic development depends on the organizational form and task. The ED specialist must be attuned to the circumstances of the community and work with laypersons as well as professionals. The ED specialist role requires him/her to utilize the expertise of the clients as well as the profession to fashion economic development alternatives for the community.

The Professional Roles

The tasks of the ED specialist are also related to the three work roles of consultant, enabler, or community organizer. Sometimes, the circumstances prescribe the role of the ED professional; more often, the organization's mission defines it.

The *consultant* role does not refer to a professional occupation but the mode of delivery. As consultant, the ED specialist provides expertise and problem-solving skills in the situation. The consultant acts as the provider of accurate technical information, showing the options available to the decision-making group.

The *enabler* is essentially a facilitator. In this role, the ED specialist is a catalytic leader who focuses on bringing people together and providing a structure for resolving community economic development issues. The enabler may also mobilize resources but seldom acts as the sole expert. Moreover, the enabler will attempt to create a continuing problem-solving capacity rather than addressing single economic or social issues (Luke et al. 1988).

The *community organizer* role has much more of an advocacy orientation. A community organizer is usually partisan to a specific group or area and acts as the catalyst to propel it to political and economic action. The organizer model requires a certain type of institutional structure, such as a community development corporation, to be effective in an economic development role.

Clearly, some of these roles are merged or even evolutionary. An ED specialist might start out as a community organizer and subsequently adopt the role of enabler and consultant as the group increases its power and capacity. The essential factor is that both the "process" and the "content/methods" task be delivered to the community in some reasonably responsible manner. The framework here is merely an outline of the role models related to the task function of the ED specialist, not a conclusive definition.

Economic Development Careers

There are several different career paths that the ED specialist can take. They generally take three distinct forms, which are as follows:

ED manager. Managing economic development organizations as separate or component agencies requires considerable managerial skill; the economic development manager must be an expert in process areas and have a strong background in both business and economics.

Analyst. Nearly all economic development organizations and consultant firms require individuals with very strong analytical skills, particularly in the areas of regional economic development, economics, and urban planning/business.

Neighborhood/community worker. At the neighborhood and small community levels, overall development and process skills are required; community organization and development training is especially useful when working at this level.

Neither the training nor the career choices of the ED specialist are uniform. There are now graduate programs and professional postdegree programs offered by both universities and recognized professional associations. The reader may wish to consult his/her local state university or state department of commerce or development to determine where such training is available locally.

Summary and Conclusion

Economic development is an emerging field of study and practice. Its antecedents lie in industrial promotion and attraction. The new aspects of the field have been discussed in this book. The basic orientation as articulated here is comprehensive development utilizing indigenous resources and capacity. ED specialists are both "process"—community organization, leadership, and capacity building—as well as "task"—economic and data analysis—oriented. The roles ED specialists play are related to the needs of the situation and the resources required. Individuals who want to become professional economic developers will have to pursue special training in economics, regional science, urban planning, or related disciplines.

References and Suggested Reading

Benveniste, G. 1983. *Bureaucracy.* San Francisco: Boyd & Fraser.

Blakely, E. J. 1979. *Community Development Research: Concepts, Issues and Strategies.* New York: Human Services Press.

Christensen, J., and J. Robinson. 1980. *Community Development in America.* Ames, IA: Iowa State University Press.

Levy, John. 1990. What Economic Developers Actually Do: Location Quotients versus Press Releases. *Journal of the American Planning Association* 56 (2): 153-160.

Luke, Jeffrey S., C. Veattriss, B. J. Reed, and C. Reed. 1988. *Managing Economic Development.* San Francisco: Jossey-Bass.

Rothman, J., J. Erlich, and J. Teresa. 1975. *Prompting Innovation and Change in Organizations and Communities.* New York: Wiley.

Rubin, Herbert J. 1988. Shoot Anything that Flies; Claim Anything that Falls: Conversations with Economic Development Practitioners. *Economic Development Quarterly* 2 (3): 236-251.

Vollmer, H. M., and D. Mills, eds. 1966. *Professionalization.* Englewood Cliffs, NJ: Prentice Hall.

Weiner, M. 1982. *Human Services Management.* Homewood, IL: Dorsey Press.

Appendix

*Glossary of Terms Frequently Used in
Connection with Economic Development*

Amortization: A payment plan by which the borrower reduces debt gradually, usually through payments of principal and interest.

Anchor tenant: The most important tenant in a development project whose lease is usually instrumental in securing financing for a commercial undertaking.

Articles of corporation: A legal document required by and filed with the state government in which the corporation is chartered. This document describes the purposes for which the corporation is formed and how it will be organized.

Assets: Things of value owned by a business, such as money, merchandise, machinery, buildings, and land.

Balance sheet: The balance sheet shows the assets, liabilities, and owner's equity of a business as of a specific date.

Basic employment: Associated with business activities that provide services primarily outside the area via the sale of goods and services, but whose revenue is directed to the local area in the form of wages and payments to local suppliers.

Bond rating: An estimate of the creditworthiness of bonds issued by a governmental unit or corporation.

Bonds: Interest-bearing certificates of debt issued by a municipal governmental body or a private corporation to finance physical improvements.

Building code: A system of uniform building regulations within a municipality, established by ordinance or law.

Business development grants: Funds provided to economic development groups by local or state governments that carry no obligation for repayment.

Bylaws: Rules, regulations, and controls set by the board of a corporation for the conduct of its business.

Capital costs: The costs a business pays for major physical improvements, such as buildings, equipment, and machinery.

Capitalization rate: The rate of interest that is considered a reasonable return on an investment.

Capitalize: To supply a project or business with funds invested by the owners or developers as distinct from borrowed funds.

Cash flow: A statement showing actual or projected cash receipts and disbursements for a specific period of time.

Central business district (CBD): Generally refers to the business area of a city or town.

Code enforcement: The power of a municipality or agency of local government to require that all properties meet certain standards of construction, maintenance, health, and safety. If a property falls below the minimum requirements and the owner does not satisfactorily repair the property, it can be declared a public nuisance and be condemned.

Commercial bank: A financial institution that acts as a savings depository and provides short- and long-term loans to business and industry as well as consumers for a variety of purposes.

Commission: A lawfully authorized group of citizens who perform certain tasks or duties in the public interest. An *economic development commission* is a good example of citizens who have been appointed to the task of developing strategies to improve the local economic base.

Community-based organization: A group representative of a significant part of a community/neighborhood that provides services that focus on community development.

Community development corporation (CDC): Usually a nonprofit organization controlled by residents of low- to moderate-income areas to help stimulate economic and physical improvement of the community.

Comprehensive approach: A means of viewing a complicated project as a total picture and then identifying all of the parts that will be needed to complete the proposed project.

Condemnation: The determination by a public authority or agency that a certain property is unfit for use.

Contractor: A person or organization performing certain tasks according to the specific terms of a written agreement.

Coordinated effort: The process of actually working with various levels of government, business, neighborhood organizations, unions, and residents with the intention of establishing who is responsible for certain parts of a project.

Corporation: An independent entity that has a legal life separate and distinct from the persons who own it. A corporation can enter into a contract without making its owners personally liable to perform it.

Cost-Benefit Relationship: In economics, the worth of an item is measured by the relationship between what it costs and what it produces.

Credit rating: A rating or evaluation made by a credit-reporting company based on a person's present financial condition and past credit history.

Debt: An amount of money owed to a lender (usually a financial institution).

Debt limit: A legal restriction on the amount of funds a city can borrow. This amount is normally a certain percentage of the assessed valuation of taxable property in the city.

Debt service: The amount of money required to make regular payments (usually monthly) on principal and interest loans.

Deed: A written document by which ownership of property is transferred from the seller (called the grantor) to the buyer (called the grantee).

Default reserve account: A special account created by public and quasi-public agencies in which a financial institution agrees to make high-risk loans and draw on this default account to recover possible losses.

Demographic information: This refers to information about the community in terms of the number of residents in a certain area, educational level, unemployment rate, racial composition, crime rates, and so on.

Depreciation: A loss or decrease in the value of a piece of property due to age, wear and tear, or deterioration of surrounding properties.

Development authority: An independent agency of local government that possesses special powers beyond those of city government. A public housing authority is a good example because it has the ability to issue special bonds for public housing.

Easement: The right to use land owned by another. An example of an easement is a utility right-of-way to allow power lines to cross another's property.

Economically disadvantaged: Any person who is a member of a family that receives public assistance in the form of welfare payments and has a total income that, in relation to family size, is lower than the poverty level determined by the government.

Eminent domain: The right of the government to acquire private property for public use, almost always with adequate compensation to the owner. Also known as *resumption.*

Entrepreneur: One who organizes, manages, and assumes the risks of a business or enterprise.

Equity: The amount by which the business assets exceed the business liabilities is known as the owner's equity.

Equity capital: Funds that owners have personally invested in an enterprise.

Escrow: The holding of money and/or documents by a third party until all the conditions of a contract are met.

Facade: The front or principal face of a building.

Fair market value: The worth or value of a property as estimated by a professional property appraiser. It reflects the price at which the property could be sold in a competitive market.

Finance charge: The total of all charges one must pay in order to obtain a loan.

Firm commitment: An agreement provided by a lender to make a loan to purchase a particular property. This commitment usually expires after a certain period of time.

Forbearance: The act of delaying legal action to foreclose on a mortgage that is in default.

Foreclosure: The legal process by which a lender forces payment of a loan by legally taking the property from the owner and selling the property to pay off the debt.

Fragmented effort: A noncomprehensive and generally uncoordinated effort to engage in economic development activities.

Franchise: A legal agreement by which a manufacturer or chain store grants the exclusive right to sell merchandise it produces or use the firm's name to sell merchandise.

Front-end costs: Capital required at the early stages of a development project, such as the cost of land and architects' fees.

Gross income: The total receipts of an enterprise, or the total receipts excluding expenses of operating a business.

Guaranteed loan: A loan that is guaranteed partly or fully by a specific governmental agency for the benefit of protecting a lender against possible losses.

Guaranty: A promise by one party to pay the debt of another if the party borrowing the funds fails to pay off the debt.

Hazard insurance: Insurance that protects against damage caused to property by fire, windstorm, or other common hazards. Hazard insurance is required by most lenders in an amount at least equal to the loan.

Holding costs: A term used by developers referring to the costs of owning land or property during the predevelopment stages of a project.

Income statement: The income statement (sometimes called a profit-loss statement) shows the net income or net loss for a specified period of time and how it was calculated.

Inflation: A rise in the general price level of goods and services that decreases one's purchasing power. For example, if the price of goods and services doubles, purchasing power will decrease by one-half.

Insured loan: A loan insured by a governmental agency or a private mortgage insurance company.

Interest: A charge paid for borrowed money. It is usually expressed as a certain percentage. For example, a $5,000 loan at 10% interest results in a charge of $500 for the first year for the use of this borrowed money.

Interest subsidy: A grant designed to lower the interest costs of borrowing. The subsidy either goes directly to the borrower or is paid on his or her behalf to the lender.

Investment syndicate: A group composed of investors, each of whom invests a sum of money and takes some share of ownership as well as a share of the risk. This enables each person to participate in investments that are larger than any one person could make alone.

Landbanking: The public acquisition of land and holding it in reserve for future public or private use.

Land-use planning: Planning for proper use of land, taking into account such factors as transportation and location of business, industry, and housing.

Land write-down: A reduction in the price of land to below fair market value. This land is usually sold by a public agency to help lower the costs of a redevelopment project.

Letter of credit: A document usually issued by a bank verifying that a borrower has credit for a specified amount.

Leveraging: A means of multiplying the availability of funds for economic development or community development programs by providing a certain amount of public funds with a proportionately larger amount of private funds.

Liabilities: Obligations of a business to pay debts, such as borrowed money and merchandise purchased on credit.

Lien: A claim that someone has on the property of another as security or payment for a debt. The lien stays with the property until the debt is paid off.

Local business development organization (LBDO): LBDOs are non-profit organizations that specialize in business assistance, such as help in collecting market data or preparing business plans and loan applications.

Market: The geographical area of demand for products, goods, and/or services. This could be one neighborhood or several neighborhoods or an entire district.

Market survey: This is a study of a specific area to determine its potential for supporting commercial activity. The market survey is designed to reveal information on resident shopping patterns, physical characteristics of the commercial area, and merchant business practices and to measure consumer purchasing power.

Net income: The amount remaining after all costs, expenses, and allowances for depreciation have been deducted from the gross income of a business.

Net worth: The value of the owner's interest in the business. This is the amount by which assets exceed liabilities.

One hundred percent retail location: The most important intersection in a retail district, usually where the highest volume of sales takes place.

One-stop permit system: A system set up in city government to enable businesses to arrange all city permits for business operations through one office.

Operating costs: Expenses associated with a business activity such as wages, rent, and utilities.

Option to buy: An arrangement to permit one to buy or sell something within a specified period of time, usually according to a written agreement.

Ordinance: A rule or law established by the local governing body (such as the city council) to control actions of citizens and the effects of their activities on others.

Overhead: Costs incurred in the sale of merchandise or services. These costs include labor, rent, utilities, insurance office supplies, and the like.

Parcel: A lot or tract of land, usually identified by ownership and a parcel number.

Partnership: A legal entity formed by two or more persons to do business. The partners must invest assets in or contribute services to the entity and must share in both the profits and the losses from the business.

Prime rate: The interest rate charged by commercial banks to business borrowers with the highest credit rating.

Principal: The amount of money borrowed that must be paid back. This amount is separate from interest and other finance charges paid for borrowing.

Public works: Facilities constructed for public use and enjoyment with public funds, such as ramps, highways, and sewers, in contrast to maintenance activities, such as street cleaning and painting school buildings.

Purchase agreement: A written document in which a seller agrees to sell and a buyer agrees to buy a piece of property based on certain terms and conditions acceptable to both parties.

Quasi-public agency: Usually a nonprofit corporation with a privately appointed board of directors whose purpose is to assist governmental agencies and the private sector to improve the general living standards of citizens more effectively. The majority of funds for such activities come from public agencies.

Rate holiday: Tax reductions given for a specified period.

Redevelopment: The physical and economic revitalization of a neighborhood or community, usually with large amounts of public funds.

Redlining: Also known as *disinvestment,* redlining is the practice by some financial institutions of designating older or declining neighborhoods or areas within a city as too risky for loans.

Refinancing: The process of paying off one loan with the money from another loan. Normally, a lender will loan up to a maximum of 90% of appraised value on residential properties and 80% of appraised value on commercial properties.

Regional investment corporation (RIC): An investment company formed by a group of citizens and businesspeople to help finance specific small businesses in a specific area.

Rehabilitation: The physical improvement of an existing residential, commercial, or industrial building.

Restriction: A legal limitation on the use of property in a deed. A deed restriction may state, for example, that the owner cannot demolish the existing building without adequate replacement because the building is security for the loan.

Retail business: A business in which the owner buys products or goods from a wholesaler and sells them to customers who make personal use of what they buy.

Revolving loan fund: Usually a municipally sponsored loan program in which a specific amount of public funds is set aside to make loans for specific purposes. As loans are repaid, the funds are loaned out again.

Right-of-way: An easement on property, where the property owner gives another person the right to pass over his or her land or allows structures to be placed on the land, such as utility poles.

Savings/thrift society: A financial institution that specializes in providing long-term housing mortgages rather than short-term business and personal loans.

Secured loan: A loan for which collateral is pledged as security.

Service corporation of retired executives (SCORE): Management assistance for small businesses provided by retired businesspeople who are identified and registered by the Small Business Administration to provide such services.

Sinking fund: A fund to which monthly or quarterly contributions are made, usually for the purpose of replacing certain assets such as machinery or for the payment of a future obligation.

Small Business Administration (SBA): An agency of state governments designed to assist small businesses with more flexible financing and less restrictive lending requirements than commercial banks.

Subsidized housing: Residential housing constructed with financial assistance from a governmental or charitable institution, or residential housing where part of the monthly rent is paid by someone other than the tenant.

Tax abatement: A reduction in property taxes for a specific property over a certain period of time.

Term loans: Bank loans generally made for periods from 1 to 5 years. These loans are designed to fit the particular needs of each borrower.

Trade-exposed: With export potential.

Venture capital: Capital subject to considerable risk and uncertainty, such as a business loan to a manufacturer of water beds when the market for such products may be at its peak.

Zoning: The power of local government to regulate the use of private property for the benefit of the entire community.

Author Index

Subject Index

About the Author

Edward J. Blakely is currently Professor and formerly Chair of the Department of City and Regional Planning at the University of California, Berkeley. As a teacher, as well as an active practitioner in economic development, he has been directly involved in all aspects of planning, project development, and finance for more than 30 years. He has been responsible for the successful design and implementation of numerous land use planning and community development projects at the regional and local level in the United States, Europe, and Australia. He has led teams in major strategic planning efforts in both the public and private sectors on both international and domestic local economic development strategies.

Blakely has most recently served as adviser and consultant to several State of California counties and cities in economic and land use development planning. In addition, he serves on a number of task forces and commissions on economic development. He is currently leading consultant projects on high-technology developments in Taiwan, Japan, Australia, and the United States. He is a leader in designing economic analysis on military base installations. He is the current President of the Pacific Rim Council on Urban Development, the leading urban planning and economic development association in the Asian Pacific. He and William Goldsmith won the 1993 Paul Davidott Award for their book *Separate Societies*.

Blakely received his B.A. in History, Political Science, and Economics from the University of California at Riverside, his M.A. in Latin American History from UC at Berkeley, and his Ed.D.

in Education, Urban Planning, and Management in 1970 from UCLA.